Gerald Pöch

Combined Effects of Drugs and Toxic Agents

Modern Evaluation in Theory and Practice

Springer-Verlag Wien New York

Dr. Gerald Pöch
Institut für Pharmakologie und Toxikologie
Universität Graz, Graz, Austria

Printing was supported by the
Fonds zur Förderung der wissenschaftlichen Forschung

Printed on acid-free paper

With 53 Figures

Library of Congress Cataloging-in-Publication Data

Pöch, Gerald, 1932—
    Combined effects of drugs and toxic agents. Modern evaluation in
theory and practice / Gerald Pöch.
        p.   cm.
    Includes bibliographical references and index.
    ISBN-13: 978-3-211-82434-4   e-ISBN-13: 978-3-7091-9276-4
    DOI: 10.1007/978-3-7091-9276-4
    1. Drug interactions.   I. Title.
    RM302.P63   1993   92-43470
    615' .7045—dc20   CIP

ISBN-13: 978-3-211-82434-4

Preface

The evaluation of effects in combination (combined effects) poses many problems to the investigator, resulting in a highly unsatisfactory situation. Different methods of analysis, contradictory interpretations of combined effects, and the different use of terms for its characterization are frustrating. Despite many attempts to overcome the discrepancies, the dilemma still exists, and we may wonder why.

Obviously, discrepancies are due in part to the differences in terminology, e.g., in the use of the same term for phenomena as well as for mechanisms, like "additivity", "synergism" and "antagonism". Also, different models are used for describing and explaining experimental results, mainly the model of "additivity" and "independence". The situation is also unsatisfactory with respect to current methods and procedures, especially with respect to time expenditure and costs to establish the widely used "isobologram".

Despite many recent publications on combined effects which deal with theory and/or practice of their evaluation, we can see that they are still based on old views and procedures which have been developed many decades ago. As a matter of fact, research on combined effects has hardly recognized the research on effects of drugs and chemicals acting alone. The latter has increasingly been based on dose-response curves (DRCs), whereas combined effects still are mainly analyzed on the basis of isobolograms.

One of the models developed for combined effects is based on the concept that drugs may act alike and produce the combined effect of a sham combination, termed "additivity". Modern understanding of the action of drugs was not available at the time the concept and model of additivity was developed and has not yet been considered (adequately). If we apply modern knowledge to this model, we recognize that agents acting by this mechanism have to bind to the same molecular site, e.g., a receptor, for which the molecules compete. Hence, the mechanism of additivity is due to a special type of competitive interaction. Recent research on binding and on receptors implies that such an interaction occurs rarely in pharmacology, and even less often in toxicology. Hence, from a mechanistic point of view, it is hardly necessary or appropriate to apply the model of additivity to combined effects, but conventional isobolograms are still based upon the "similar" action.

The other model, in which chemical substances as well as physical factors independently lead to the same phenomenological effect, does not require understanding of molecular pharmacology and toxicology. It is less popular than the additivity model but deserves more attention. By comparing experimental results with those expected for independent effects, we can see whether or not the results can be explained by independent action. Since the model of independence in action implies that one agent is not affected by the other, it is the proper reference for enhanced and diminished effects, for potentiation and antagonism.

Although for certain investigations some procedures appear more appropriate than others, DRCs, in general, will provide most of the information we want to know. Of these, the DRCs of one substance alone and in the presence of a fixed dose of another substance will often provide enough information with a minimum of costs and time expenditure. This approach offers a reduction in time and costs for studies of combined effects by direct analysis of DRCs. As with antagonists in pharmacology, a DRC of a drug A is tested in the absence and presence of a fixed dose of a drug B, acting in the same direction as A. The observed effects can basically be compared with two theoretical DRCs, one for additivity, one for independent effects. Comparison with additivity is of value where we are interested whether or not drugs share the same molecular binding site of action. Such situations are rare. Comparison with independence allows differentiation between potentiation and antagonism and its appropriate quantitation.

The new approach rests on modern views on binding and action of drugs and on a careful study of established principles and methods with respect to their assumptions and practicability. The reader is led step by step to a meaningful and practical evaluation of combined effects.

Of course, we have to consider that there are many ways to look at and many procedures to evaluate combined effects. As long as the appropriate procedure is used, each of these offers some advantages and disadvantages. It depends on what we want to know. Also, we have to keep in mind that a certain phenomenon can be caused by different mechanisms. The reader will understand that it is extremely important not to mix phenomena and mechanisms. Besides presenting a new and modern approach to the evaluation, this book is aimed at a balanced, yet critical, description and discussion of existing methods.

The described and discussed methods are not restricted to DRCs but also include time-course and single-dose studies. The latter procedures are mainly based on the conclusions derived from DRC-studies and on generally accepted principles, respectively, rather than on DRCs directly. Some pragmatic conventional procedures, like fixed-ratio combinations (mixtures) are also described and discussed. Tips and hints are given in many chapters for practical evaluation, e.g., of literature reports.

Chapter 1 starts with the basic principles of actions and interactions, including "synergism" and "antagonism" in situations in which the DRCs of one drug A are tested in the presence of a fixed dose of a drug B, which exhibits no effect by itself. Chapter 2 deals with the models of additivity and independence, and the site-directed analysis. It will be shown how the

DRCs of additivity and independence look like in various graphical presentations, e.g., in linear-dose log-response curves. Comparison with dose-additive combinations primarily give important hints to the sites of action of substances tested in combination, whereby the additional comparison with independence may strengthen or weaken the conclusions based on comparison with additivity. In addition, it describes the relationship between dose-additive and independent combinations, their dependence on the steepness of DRCs having been realized only recently.

Chapter 3 describes the different opinions and the different use of the terms synergism (or potentiation) and antagonism. It deals with various procedures for quantifying effects in combination, including a new graphical plot for illustration of enhanced effects. The suggestion to use independent effects as the reference for potentiation (synergism) and antagonism is substantiated. This chapter also refers to the combination index, derived from the doses for a given effect. It is inappropriately used by many investigators to express the magnitude of combined effects, because of the different relationship between increases or decreases in doses and changes in effects.

Chapter 4 describes and illustrates the new approach from the methodological point of view, aimed at a procedure that can save time and expenditure, yet can provide important clues to the site of action of drugs as well as to the magnitude of combined effects. The latter is immediately evident by the actual effects rather than indirectly derived from the doses necessary to obtain the same effect.

Chapter 5 describes and illustrates the evaluation of time-course studies from a basic and practical point of view, followed by a comparison of studies with DRCs and time course. In Chap. 6 single-dose and other studies, and in Chap. 7 a new graphical presentation of enhanced or diminished effects in combination, the combined-effect graph, and other graphs are described.

Chapter 8 provides numerous experimental examples in many fields, ranging from biochemistry to herbicide research. It includes examples which have not been published so far or have not been analyzed in the way presented. This chapter will convince sceptical readers that the new approach offers a number of advantages over conventional procedures. It also compares, on empirical grounds, observed effects in combination with independent effects.

Chapter 9 describes and explains the pitfalls of the isobologram approach and the weakness of the Chou–Talalay evaluation by the median effect plot together with isobolograms. Chapter 10 compares the new with the conventional approach, and hints are given as to how to handle problems of evaluation of combined effects in the future.

In an appendix, a "Guide to Practical Work" in the evaluation of DRCs is given with two examples described in great detail, followed by additional two examples for practising. The reader should in the end be able to plan, analyze, and evaluate combined effects in a more satisfactory manner than by conventional practice.

All the figures were especially drawn or adapted for this book, in a manner that should make it easy to understand the text and to compare various graphs with each other.

The valuable help of Mrs. Betty Oberer and Mrs. Christa Kern for drawing the figures and typing the tables is greatfully acknowledged. I also thank many colleagues for critically reading the drafts of the chapters, and for stimulating and valuable discussions, especially Peter Dittrich, Hans-Dieter Unkelbach, Jürgen Sühnel, and Klaus Groschner. My special thanks go to Rod Reiffenstein who carefully reviewed and checked this manuscript. Finally, I would like to thank Springer-Verlag for their willingness to accept my wishes and ideas for the publication of this book.

Graz, December 1992   G. Pöch

Contents

Contents

1 Principle considerations of drug actions and interactions

1.1 Introduction

Many investigators have to face the question of how to design an experiment and evaluate combined effects, for various reasons and different goals. Some may be confronted with this problem by coincidence, others may have an interest in this field from the beginning. Which of the many approaches one follows, may depend on the special situation and its circumstances, like whom one asks for advice, which textbook or paper is consulted, etc. The longer one is concerned with combined effects, the more one realizes that almost everyone has their own opinion. Probably the great majority of researchers shares the opinion that the topic of combined effects is a continuing problem, especially with "synergistic" agents in combination.

In the opinion of the author, one reason for this dilemma is the neglection of modern research on the action of substances, especially the *binding* to receptors or, more generally, to binding sites. At the time the "additivity" concept was developed, we only had a concept, a fiction of "receptors" (see Ariëns 1979, Greaves 1976). "With the possibility of receptor localisation, identification and isolation, the use of receptor-binding studies ..., the receptor concept has been converted to hard reality" (Ariëns 1979). Receptors not only are binding sites for drugs (see Bourne and Roberts 1987) but also for chemicals which exert toxic actions (see Klaassen and Eaton 1991), and it appears that almost all substances, including xenobiotics (Hinson and Roberts 1992), bind to certain molecular structures (sites) located on cell membranes and on intracellular macromolecules. Also, molecules of different compounds which bind to the same molecular site *interact* by competing for it when applied together.

Let us now look at combined effects from two sides, from one, to study, predict, and describe *phenomena*, from the other one, to obtain information about the *mechanism* of action and interaction of compounds. (The term "interaction" is used in this book to indicate a true *inter*action or simply a combined action.) We would expect that those who are in favour of the phenomenological point of view, will use descriptive terms, only. This is not the case. One of those respected researchers is the late Morris Berenbaum, who has recently published a review article (Berenbaum 1989). He used the mechanistic term *"non-interactive"* to describe the phenomenon of a dose-additive combined effect. The term "no interaction" is used by others to characterize combined effects due to independent actions of the agents under study (Elashoff et al. 1987), which is considered appropriate by the author of this book, for reasons given later. Still others use the same term, "no interaction", to characterize a situation in which the combination is just as effective as one of its constituents used on its own (Calamari and Alabaster 1980).

Let us compare experimental results of combined effects of "synergists" with *mechanistic models* just as it is done with antagonists or inhibitors (e.g., Bourne and Roberts 1987). For instance, experimental results are compared with the model of competitive interaction. In this example, when results do not conform to the model of competitive interaction, they cannot be explained by the mechanism on which this model is based. However, it is important to realize that when results (appear to) conform to this model, a competitive interaction is not proven. It may be observed that the dose-response curve (DRC) of a receptor agonist is shifted in parallel to the right by a given antagonist, as the model of competitive interaction predicts. However, this finding has first to be seen as phenomenon that theoretically could be brought about by competitive interaction but by other mechanisms as well. For instance, if this study is repeated with a wider range of antagonist concentrations, a deviation from the model of competitive interaction may well be evident. In other words, and in a more general way, only a significant deviation from results to be expected by a certain mechanism, excludes that the agents studied interact in the expected way. On the other hand, however, lack of significant deviation from a model does not prove that this particular mechanism explains the observed phenomenon.

Another point is that different models can be used to ask *different questions*, irrespective of whether they are purely descriptive or based on mechanisms. As referred to above, even the comparison with mechanistic models requires that we first look at experimental results from a phenomenological point of view. Non-mechanistic models, like effect-summation, do not allow (direct) conclusions on mechanisms of actions or interactions.

For instance, the "model" of combined effects conforming to the *sum of individual effects*, i.e., to the summation of effects, is descriptive in nature. It does not allow direct conclusions about mechanisms of interaction since it is not based upon a mechanism of interaction. However, when the combined effects observed are greater than the sum of individual effects, they must be greater than effects of an independent interaction. Independent effects are less than the sum of individual effects; they come close to the sum of small individual effects but are much lower at higher individual effects, as will be shown in Chap. 2.4.2. It should be noted that the sum of effects is sometimes described as additive, to be understood as effect-additive. The model of independence as well as the model of additivity (of doses) will be the topic of the forthcoming chapter.

This introductory overview of some of the many confusing aspects of combined effects is intended to alert the reader and to shed light on the controversy among researchers in the evaluation of combined effects. Before we take a closer look at actions and interactions of chemical agents or physical factors, let us consider a few important points. We have already touched the problem of phenomenon and mechanism.

Another point of interest is to use the *appropriate experimental procedure and evaluation* to get an answer to a certain question. As a matter of fact, the different models of combined effects allow certain appropriate questions but are of less or no value if we ask other questions. For instance, the additivity model considers two or more *substances*, not physical factors, to behave as dilutions of one and the same substance. Comparison of effects observed in combination with this model may provide hints as to the molecular site of action of drugs. However, this model is often inappropriately used, i.e., to answer the question of whether or not effects in combination are enhanced or diminished. This point will be dealt with in the following chapters, especially in Chap. 9.

For many investigators, like clinicians or toxicologists, the *extent* of a combined effect is of great interest. The question is, how we should express combined effects and deliver such information. Here, only an outline for this approach is drawn. The best information we can get from studies on effects in combination is to indicate the combined effect as well as the individual effect of the components rather than to state that the effects in combination are greater than "additive", for instance. A proper way to express the results of a dose-response study by DRCs is to indicate the shift in the DRC of a drug A which is observed in the presence of the drug B. However, we should give additional information regarding the effect of A (steepness of DRC) as well as B (effect, when applied singly).

With regard to enhancement, there are also problems and questions related to adequate procedures of *presentation* of results. Many of them do not give an immediate impression of whether the effects of a compound are enhanced or diminished in combination. For instance, some publications deal with the effects of *mixtures* of drugs A and B, where a DRC to A and B alone and in a fixed-dose *ratio* is presented on a common dose scale for A, for B, and for the total dose of A plus B. It is almost impossible in certain cases to see that the effects of A and B in combination are enhanced or diminished.

A more common problem is the interpretation of the *isobologram*. Although it appears that this graph presents a clear picture of an action of two agents in combination, where A and B act phenomenologically alike, only the *doses* of A and B for equal effects in combination are given rather than changes in effects. The latter are inappropriately concluded from changes in equieffective doses, since additive combinations, and therefore "overadditive" and "underadditive" combinations represent different degrees of enhanced effects in combination – depending on the steepness of DRCs (Unkelbach and Pöch 1988). A critical look at isobolograms will be taken in Chap. 9.

It is noteworthy to mention the *different use of the same term* as well as *different terms for the same phenomenon and/or mechanism*. It appears that this confusing practice often arises from mixing phenomena and mechanisms. For instance, some investigators use the term "synergism" to indicate a phenomenon, others to characterize a mechanism of interaction. In the following part of this as well as in other chapters, great effort has been undertaken to clearly separate phenomena and mechanisms.

The effects of substances or physical factors depend on the dose of substances or the intensity of factors. Since the effects develop within a certain time, we can either look at the dose response at a given time or we can study the time course of a given dose.

1.2 Dose-response curves

The dose-response relationship of a given compound is probably best known for endogenous substances or for many drugs, e.g., interacting with endogenous transmitters, but can be applied to other agents as well. The dose-response relationship provides information which we need for a deeper understanding of combined effects. DRCs are the graphical expression of the dose-response relationship, whose inherent information we can directly exploit. Also, we can construct theoretical DRCs for model interactions.

Let us start to describe and discuss DRCs, illustrated in Fig. 1a and b. Figure 1a and b schematically shows the increase in response with increasing doses of a substance A. The dose of A above a certain threshold exerts some kind of response or reaction of cells etc. At lower doses no effect can be observed, termed E_{min}, the minimum of an effect E. Increasing doses exert increasing effects up to a maximum, E_{max}. Besides the characterization of the dose response with respect to the effect, a DRC is also characterized by the doses necessary to produce certain effects, e.g., by the half maximum effective dose, the ED_{50}. Furthermore, the relationship between increases in effects with increasing doses varies with different steepness of DRCs, mathematically expressed, as "slope".

1.2.1 The parameters of dose-response curves

Four parameters have been mentioned, with which a DRC can appropriately be described: E_{min}, E_{max}, the ED_{50}, and slope.

E_{min} and E_{max}
In some areas the threshold dose, and hence E_{min} is of great interest, in others, this parameter is identical with the control value. E_{max} indicates the maximum response which can be obtained or observed with a drug or factor. It is of interest to note that drugs which bind to the same molecular site, e.g., a receptor, can exhibit marked differences in E_{max}. This is the reason why such drugs can function as agonists, partial agonists or antagonists.

The dose-response relationship has been deduced mathematically from binding of an "agonist" to a molecular site (receptor) which then triggers events leading to an "effect" (see Ruffolo 1983). Although we still

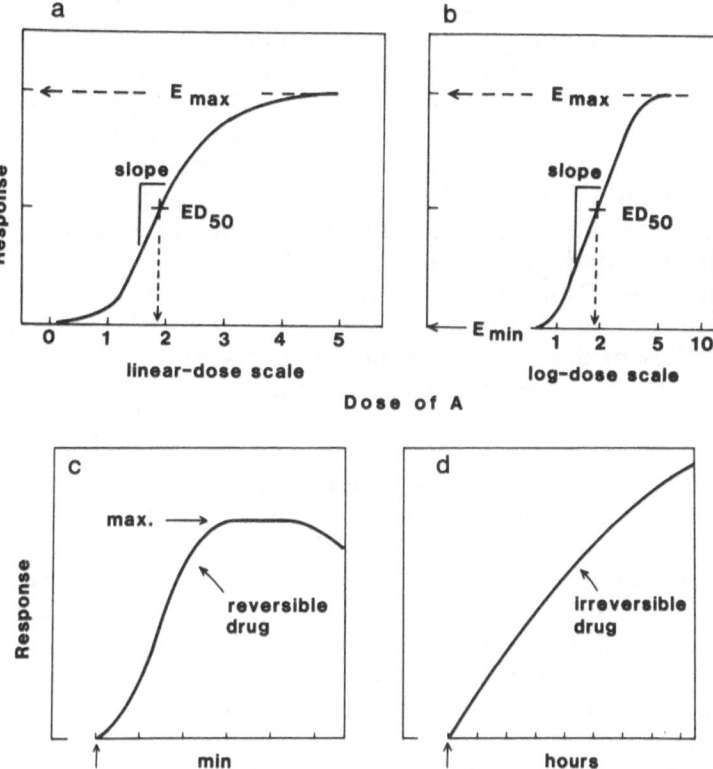

Fig. 1. Characteristics of dose-response curves (DRCs) and time-course studies. E_{min}, E_{max}, ED_{50}, and steepness (slope), schematically shown for linear-dose response curves (**a**) and for log-dose response curves (**b**). E_{min} is indicated in **b** only. Time course of response to reversible (**c**) and to irreversible drugs (**d**), schematically

do not fully understand coupling between binding of agonists to a receptor and the following effect, we can assume that, in principle, DRCs of agents other than receptor agonists also have to bind to certain structures in order to somehow elicit a response.

A DRC can adequately be described by a four parameter logistic equation (De Lean et al. 1978, 1988)

$$Y = [(a - d) / (1 + (x/c)^b)] + d$$

with the four parameters: $E_{max} = a$, slope $= b$, $ED_{50} = c$, $E_{min} = d$. Y is the response, $x =$ dose.

ED_{50}

The ED_{50} is the dose or concentration at which 50% of the maximum effect or response is attained, i.e., at

$$[(E_{min} + E_{max})] / 2.$$

The ED_{50} corresponds to 50% when $E_{min} = 0$ and $E_{max} = 100\%$, e.g., in dose-frequency studies; in other situations it does not, e.g., when $E_{min} = 20\%$ and $E_{max} = 100\%$. In the latter example, the half maximum effect is $(20 + 100) / 2 = 60\%$ at which the half-maximum effective dose, the ED_{50}, has to be determined. This is the way pharmacologists usually determine the ED_{50} (e.g., De Lean et al. 1988; see also Chap. 2.3.2). Others consider the ED_{50} as the dose which actually produces 50% effect. The latter procedure is used by those who are interested in the doses which produce 50% effect, regardless of E_{min} or E_{max}, e.g., to construct an "ED_{50}-isobologram" (e.g., Brunet et al. 1986, Pancheva 1991). In isobolograms equieffective doses determine lines of equal effects, the isoboles, therefore the ED_{50}s must by necessity represent doses which in all combinations produce a fixed response, i.e., 50% effect (see also Chap. 2.1.2).

Slope of DRCs

The steepness of a DRC can be expressed by the "slope" parameter. Looking at DRCs of different agents shows different values of slope. It is generally accepted that a slope of "unity", i.e., slope = 1, is expected in the case where receptor agonists or enzyme substrates bind to a binding site which is not composed of subunits. The steepness of a DRC of A, and of B is a most interesting aspect in the evaluation of combined effects, especially where slope values deviate from 1. Values > 1 (or < 1) can be caused by two basic mechanism, one involving cooperativity at the binding site, another one on a functional basis (Katzung 1987).

Binding sites can be composed of protomers, or identical subunits, of an oligomeric protein (see Wong 1975). Positive cooperation is characterized by slope values > 1, and is explained by an interaction in which the binding of one molecule of a substance to one subunit allosterically enhances binding to the remaining subunits of a catalytic site of an enzyme (see Wong 1975) or of a receptorprotein (see De Lean and Rodbard 1979). Empirically, DRCs of receptor agonists often show slightly greater slope values than 1 with many possible underlying mechanisms (Ariëns et al. 1964a, 1979). The DRCs of cholinergic agonists at the nicotinic receptor were reported to show a slope between 1.5 and 2.6 (Peper et al. 1982). This receptor appears to be composed of five subunits, two of which are identical (see Peper et al. 1982, Raftery et al. 1984, Guy and Hucho 1987).

Let us now turn to functional "cooperativity". Many, including some clinical responses, are characterized by (very) steep DRCs "which could result from cooperative interactions of several different actions of a drug (e.g., effects on brain, heart, and peripheral vessels, all contributing to lowering of blood pressure)" (Katzung 1987). Examples of DRCs with different slopes are given in later sections of this book.

In general, DRCs with slope values < 1 are deemed flat DRCs, > 1 are considered steep, >> 1 as very steep. To aid understanding of slope-related problems in the evaluation of combined effects, DRCs are designated in this book as *flat DRCs*, *normal DRCs*, *steep DRCs*, and *very steep DRCs*. They refer to DRCs with slope values of < 1, around 1, 1.6, and > 1.6, respectively.

Translated into effects at the twofold ED_{50}, we find < 66.6% with *flat* DRCs, ≈ 66.6% with *normal* DRCs, 75% with *steep* DRCs, and > 75% effect with *very steep* curves.

It will be shown that drugs other than transmitter-like agents typically exhibit normal DRCs, i.e., slope around 1, or slightly higher, whereas anesthetics or cytostatics typically show steep to very steep DRCs (slope values around 1.6 or above). Most toxic agents exhibit very steep DRCs with slopes between about 2 and 10.

Dose scales

The doses or concentrations can be plotted either on a linear-dose scale (Fig. 1a) or on a log-dose scale (Fig. 1b). Note that all DRCs plotted on a *log-dose scale* are characterized by S-shaped curves, including *exponential DRCs* (Unkelbach and Pöch 1988). The latter all have the same slope (1.6) and can be linearized when the doses are plotted on a linear and the response on a log scale (Berenbaum 1981). Log-dose scales are frequently more appropriate than linear dose scales, especially with agents where the minimum and maximum effective doses span over a wide range. On a *linear dose scale*, only DRCs with slope values higher than 1 are S-shaped; the DRCs with slope = 1 are hyperbolic curves (see Wong 1975, Chou and Talalay 1984).

1.2.2 Models and curve-fitting

Various models are used to describe and to fit experimental data points to curves, including models for cooperative interactions (see Wong 1975, De Lean and Rodbard 1979, De Lean et al. 1978, Meddings et al. 1989). Fitting curves to data using nonlinear regression was recently considered superior to linear regression of transformed data (Motulsky and Ransnas 1987). Most of the DRCs shown in this book were obtained by fitting data points to nonlinear curves by the program ALLFIT which utilizes a logistic function (De Lean et al. 1978, 1988).

1.2.3 Comparison of curves

When two or more substances are tested alone and in combination it appears of interest to interpret and compare the parameters, not the least the slope values. There is general agreement that agents which bind to the same site should exhibit the same slope, whereas E_{max} may differ, as mentioned. With drugs which all reach the possible maximum in response, their DRCs have then to be parallel (Levine 1983). Note, however, that the same slope does *not* exclude different binding sites. Not much is known about the mechanism of steeper DRCs with slope (much) greater than 1, except with respect to cooperative *binding* (De Lean and Rodbard 1979). Cooperative interactions on a *functional* basis were mentioned above. They are likely to be associated with binding to more than a single type of binding site.

1.3 Time course of effects

DRCs of reversibly acting drugs are usually based on the respective maximum effect achieved within a specific time, whereas DRCs of irreversibly acting compounds show the response at a selected time. Figure 1 c and d schematically illustrates the development of effects with time. Reversibly acting drugs usually reach a maximum within minutes which lasts for a certain time, after which the response declines (Fig. 1 c), due to a decrease of the active drug molecules at the "receptor", e.g., by metabolism, or caused by a decrease of the response itself, sometimes called desensitization. Irreversibly acting compounds show an increase in effect with time, usually for hours, with no decline for a long period of time (Fig. 1 d).

1.4 Phenomena and mechanisms: interactions between drugs – antagonism and synergism

Numerous possibilities and types of interactions between drugs can occur, e.g., resulting in the phenomena of synergism or antagonism. For instance, the effects of a drug A can be antagonized by a drug B in a competitive fashion or in a manner which is different from a competitive interaction. The following section deals with the phenomena of synergism and antagonism; their interpretation from a basic point of view is dealt with in detail in Chap. 3.

One can find very different definitions for antagonism and synergism for simple and complex interactions and of complex interactions. The term "simple interactions" here describes the interaction between an active and an inactive drug, the term "complex interactions" characterizes interactions between two active drugs, e.g., both acting in the same direction.

It might be useful to look for meaningful definitions based on changes in effects. A dictionary of pharmacology (Bowman et al. 1986) defines an antagonistic drug as "a drug that counteracts or prevents the action of another drug or endogenous body chemical". A "synergistic action" is there defined as "a drug action that increases the effect produced by another drug". It is likely that the majority of pharmacologists and other researchers will accept these definitions. However, especially in interpretating complex interactions, many of them do in fact deviate from these definitions by comparing observed effects with "expected effects", but there is no general agreement what constitutes "expected effects".

According to the above definition, an inactive drug B, which exhibits no effect by itself (in the applied dose) will diminish the effects of A in antagonism, and will enhance them in synergism. Mathematically,

$$E_A + B < E_A$$

describes antagonism,

$$E_A + B > E_A$$

characterizes synergism, where E_A is the effect of a drug A, and B is the antagonist or synergist, respectively. This definition is not restricted to any particular situation. Therefore it should in principle apply to single-dose, time-course, and dose-response studies. The latter provides most of the information researchers are usually looking for. We will therefore study DRCs of A in the presence of B, in order to see how one drug affects the action of another drug.

1.4.1 Simple interactions

For convenience, those interactions are designated "simple interactions" in which one component (B) by itself does not show an effect. The phenomenon of antagonism ($E_A + B < E_A$) is characterized by decreased effects of A at certain doses, due to B. Two types of simple antagonistic interactions can be differentiated, depending on how the DRCs of A are affected by B. They are illustrated in Fig. 2a and b. Submaximum but not maximum effects of A are decreased by B in Fig. 2a. Here, the DRC of A is shifted to the right by a drug B, i.e., we observe an increase in the ED_{50}. This means that a higher dose of A is necessary in the presence of B to exert a half maximum effect.

This type of interaction can often be explained by *competitive* interaction but other mechanisms can also show this phenomenon, e.g., allosteric interaction. Competitive antagonism requires that the interaction follows the mass action law, explaining competition between binding of the agonist (A) and the antagonist (B) (see Van den Brink 1977, Kenakin 1987). This is usually proven by displacement of binding curves or by Schild-plots (Arunlakshana and Schild 1959), analyzed with respect to the mass action law. The antagonistic action of a competitive antagonist B is explained by its inability to activate, upon binding, the same chain of

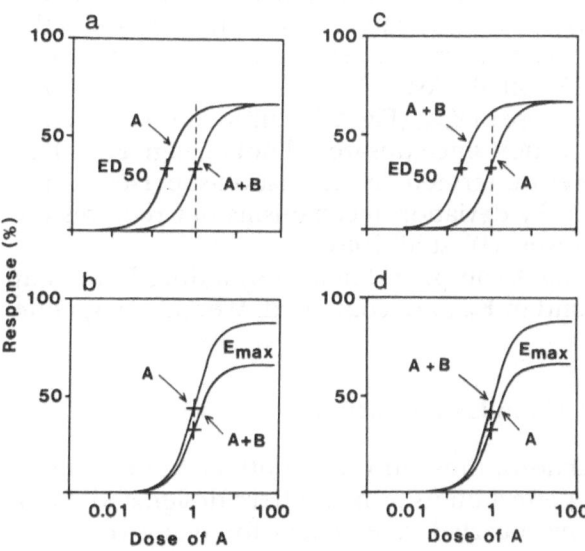

Fig. 2. Prototypes of antagonism (**a, b**) and synergism (**c, d**), schematically. The DRC of A is shifted to the left (**a**) or to the right (**c**) with corresponding changes in the ED_{50} (+) but with no change in E_{max} (**a, c**). The DRCs of A are shifted upwards (**b**) or downwards (**d**) with corresponding changes in E_{max} but with no changes in the ED_{50} (+)

reaction as the agonist A. However, B decreases the chance of A to bind to the receptor and thereby exerts an antagonistic "effect" which, in turn, is dependent on the concentration of the agonist and the antagonist at the receptor. The competition between agonist A and certain concentrations of the antagonist B explains the magnitude of the shift of the DRC of A to the right. This type of interaction has been modelled by pharmacologists, and numerous examples of competitive agonist-antagonist interactions in functional studies have been published in pharmacology journals, which are in line with this model (see also Kenakin 1987). Furthermore, functional studies, e.g., with isolated smooth muscles, are in line with binding studies, showing that there is a predictive competition for binding between agonists and antagonists (e.g., Limbird 1986). In binding studies, such a competition can directly be demonstrated between various antagonists for binding to the same site, i.e., with various ligands (e.g., Tallarida et al. 1988).

Another type of antagonism is characterized by a decrease in E_{max} (Fig. 2b). In this case, the maximum effect of A is depressed by B, and this antagonistic effect of B cannot be reversed by an increase in the dose of A. Again, there is a typical mechanism causing this phenomenon, the *noncompetitive* type of interaction (see Van den Brink 1977). This type of antagonism is explained by an inhibitory action of B beyond the receptor, activated by A.

Where, in the opposite direction to antagonism, the effects of A are enhanced by B $(E_A + B > E_A)$, we observe changes in the ED_{50} or in E_{max} mirrorlike to antagonism, as schematically shown in Fig. 2c and d. This phenomenon is generally regarded as "synergism", "potentiation", or "supersensitivity". This means that in synergism we can observe a left shift or an upward shift of the DRC of A by B. The left shift (Fig. 2c) is seen where a drug B inhibits the breakdown of A to ineffective metabolites, e.g., with cholinesterase-inhibitors enhancing the effects of acetylcholine (Green et al. 1978), simply by raising (deviating) the concentration of the transmitter at its receptor. This type of interaction was therefore termed "deviation-supersensitivity" (Westfall 1981).

The observation of an increase in E_{max} (Fig. 2d) cannot be explained by an increase of A at its site of action since this would not change E_{max}. This type of interaction was therefore termed "nondeviation-supersensitivity" since it cannot be explained by deviation mechanisms but requires the assumption of other mechanisms (Westfall 1981).

Needless to say that a drug A can be influenced by a drug B in a way where changes in the ED_{50} and in E_{max} are combined. We will not go into further details here.

1.4.2 Complex interactions

In contrast to simple interactions, those in which both components in a binary combination exert an effect on its own, are here designated "complex interactions". They have puzzled researchers for many years. To understand why, we should look at the phenomena and their possible underlying mechanisms.

Phenomena

In these interactions, the parent compounds exhibit an effect which *phenomenologically* appears as the *same effect*, e.g., muscle contraction or lethality, although complex interactions can also occur between agents with opposite actions.

Considering the basic definitions referred to above, we can express antagonism and synergism in complex interactions as follows. Since here drug A and drug B by itself produces an effect, we look at common situations in which the effects of drug A and drug B, singly, are different in magnitude, and we express this difference by upper and lower case letters (E, e). $E_A + e_B$ then indicates a combination of drug A with drug B, where drug A exhibits a greater effect than drug B in the same direction.

Antagonism between A and B can then be expressed by

$$E_A + e_B < E_A \text{ and by } e_A + E_B < E_B,$$

which means that the effect of A is antagonized by B, and B is antagonized by A. Synergism is thus expressed by

$$E_A + e_B > E_A \text{ and } e_A + E_B > E_B,$$

as the antimere of antagonism. Hence, in synergism, the effect of A is increased by B, and vice versa.

We have not yet considered any model, like additivity or independence. The latter, two fundamental types of complex interactions, will be described in Chap. 2. Because of special circumstances, not yet touched, we will first approach the evaluation of complex interactions by a descriptive and by an interpretative analysis, before we begin to describe and discuss the models mentioned.

Descriptive analysis of combinations
In general, we find pure descriptions of interactions where only few data are available. For instance, we might find the description of effects of A = 50%, B = 15%, and A + B = 80% effect. On the basis of the above considerations, the combined effect appears to conform to synergism. In a more restrictive sense we may use the term *enhancement* instead of *synergism*. We could have A = 50%, B = 15%, and A + B = 30% ; this may be characterized as *diminution* or, more broadly, as *antagonism*. Such a purely descriptive analysis may appear even more naive than the above definitions of synergism and antagonism. However, the use of the terms "enhancement" and "diminution" instead of "synergism" and "antagonism" avoids misunderstandings, since the latter expressions are used by some researchers as descriptive terms, and in an interpretative manner by others.

There are several ways to describe results of combination experiments. In some situations we might simply be interested from a practical point of view whether the effect of a threshold dose of a drug A may be increased by a drug B, and how effective B alone has to be to achieve this. Likewise, we may want to know how much the effects of A at a certain effect level can be increased by B, e.g., as a function of effects of B. In such situations we could stick to purely descriptive presentations of results.

The principle of a graphical expression of a descriptive analysis, along the line mentioned above, is shown in Fig. 3. For convenience, it is termed "combined-effect graph". On the y-axis the enhancement in the effect of A in the presence of B is shown, on the x-axis the effect of B alone is indicated. Let us first look at DRCs of the example in Fig. 3a. It shows the DRC of A in the absence and presence of B, and the changes in effect (arrows) of A by B at the indicated effect level 0% and 50% . In Fig. 3b the increase in effect is shown from 0% to 25% , whereas Fig. 3c illustrates the increase in effect from 50% to 80% of the maximum. In both graphs the arrows originate from 15% effect of B. If this increase in effect from 0 to 25% and from 50 to 80% had occurred at a higher effect of B, e.g., from 20%, the arrows would have been displaced on the x-axis to 20% effect of B.

Hence, this graph shows the increase in effect of A at a selected effect level as a function of the effect of B. This trivial descriptive graph may be easily extended to a comparative analysis with the independence model; the latter will be described in the following chapter. Here, it is of interest that a diagonal line in this graph represents independent effects in combination. This combined-effect graph will be described from a practical point of view in Chap. 7, and examples will also be given in other chapters.

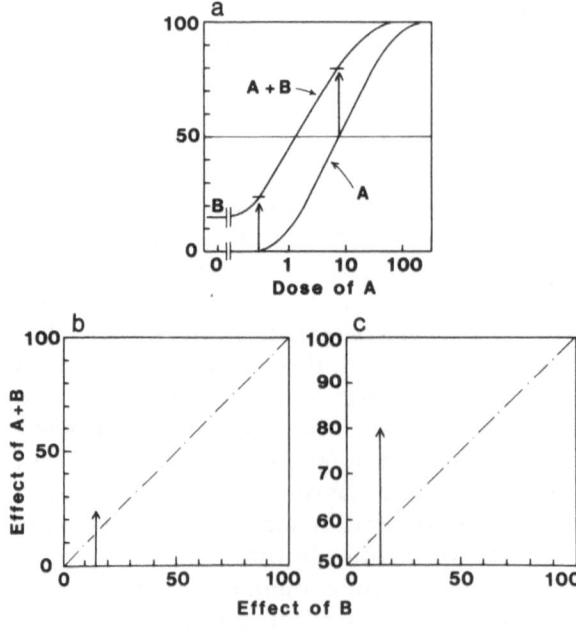

Fig. 3. Expression of enhancement of the effect of A by B. **a** DRCs of A alone and in the presence of B. Enhancement, shown by arrows, is exemplified for the threshold and 50% effect level. **b, c** Enhancement, again indicated by arrows, is shown on the y-axes as increase in effect from 0 to 25% (**b**) and from 50 to 80 % (**c**) as a function of the effect of B (15%) on the x-axes. The broken dotted line indicates the enhancement expected for an independent action of drugs A and B

Whether expressed in this or in other ways we may, as already indicated, extend a descriptive analysis to a *comparison with models*, i.e., with "expected effects". Chapter 2 deals with the models with which observed effects can be and have been compared. We can do such a comparison again by restricting our analysis to a description of whether or not the observed effects in combination agree with the effects expected – without drawing conclusions beyond this point.

Interpretative analysis of combinations
The more data points are available, the greater the temptation to check whether the results in combination could eventually be *explained* by the mechanisms on which a model is based upon. Actually, most researchers extend their analysis of combined effects to an interpretation of the findings in one or the other way. Unfortunately, it is not always clear whether a given analysis is descriptive or whether actions or interactions of drugs are interpreted with respect to mechanism underlying the observed phenomenon. This statement might appear peculiar to the many readers, however, most of the terms used in the analysis and evaluation of combined effects are either used for description of a phenomenon or for the interpretation of the underlying mechanisms (synergism, antagonism, interaction). Also, descriptive terms, like "overadditive", are used as synonyms of, or interchangeably with, "synergism" or "potentiation". This circumstance can be understood quite easily, since many scientists first describe a phenomenon and then draw conclusions for explaining it. Problems with this approach are preprogrammed in a field where a phenomenon can be analyzed and evaluated differently, leading to misunderstandings among investigators.

In order to understand such controversies in the literature it is necessary to study the phenomenon which is described, and carefully examine in which way it can correctly be interpreted and which terms and interpretations are inappropriate or misleading.

Mechanisms of interaction

In the interpretation of phenomena of antagonism and synergism we have already introduced considerations of mechanistic models. Actually, we could describe a phenomenon simply as a phenomenon, e.g., the changes in the ED_{50} and/or in E_{max}, but it is much more satisfactory to find an explanation for the phenomena. For this purpose, we can primarily develop mechanistic models, with which we can compare our observations.

As already mentioned, we can assume special mechanisms of interaction between drugs A and B, e.g., a competitive interaction at a transmitter receptor. Two types of competitive interaction are of special interest for the evaluation of complex interactions, namely "competitive synergism" and "competitive dualism" (Ariëns et al. 1956a). The latter may apppear as competitive synergism under certain conditions, which will be described in Chap. 2. It might be useful to note at this point that "competitive synergism" is a mechanism which can explain "additivity". All types of competitive interaction can be understood by the competition of molecules at the same molecular site. The outcome, i.e., the effect of A and B in combination depends on the intrinsic activity of A and B. However, not all "additive combinations" can be taken to indicate competition at a certain binding site but can occur by mere coincidence (see Chaps. 2.1.4 and 8.2.3).

2 Concepts and models of interactions – additivity and independence

This chapter deals with the two most important models of combined effects of two substances, in which the latter phenomenologically and qualitatively exhibit the same effect. These models are based on different concepts. In one model, the combined effect of A and B is explained by the concept that B behaves like a *dose* of A (and, consequently, vice versa), and in the other that the *response* to A is independent of B (and vice versa). When observed combined effects correspond to one or the other model, we can first interpret the phenomenon itself on the basis of the model. Secondly, we can draw conclusions with respect to the underlying mechanism. To aid understanding, an overview is given in Table 1, which is explained in the following sections.

Other models can be found in the literature, e.g., the "multiplicative model" or the model of "effect summation" (see Berenbaum 1989). These models can be looked upon as purely mathematical models which may or may not be connected with the models of dose-additive and independent combinations. However, reference will be given in the text where and when the latter can be described by the addition or multiplication of "effects".

For convenience, the following phrases will be used, which are used by others also. "Greater than additive" means that the combined *effects* are greater than expected for an additive combination (in DRCs). It also means that the same effect is produced by lower *doses* than expected for additive combinations (in isobolograms). Analogously, "greater than independent" is used to indicate greater combined *effects* than expected for independent action (in DRCs), and lower *doses* than expected for independently acting agents in combination (in isobolograms). "Independent effects" refer to effects of agents or factors which independently act in combination.

2.1 Concept and phenomenon of additivity

One of the most controversial terms in the topic of combined effects are "additive" and "additivity". Although many researchers refer to the addition of doses, from which the combined effects can be derived, others consider additivity in terms of effects, i.e., as effect-additive. Hence, we can find dose-additive or effect-additive combinations in the literature, both of which termed additive. Effect-additive combinations are not based on a pharmacological or toxicological mechanism of interaction. "Additivity" is either taken as a phenomenon or is considered mechanistically but its meaning is not always clearly stated in the literature. So, there are many explanations for divergent conclusions based on additivity.

Anyway, in the opinion of the author the concept and model of additivity is of limited value, contrary to the belief of respected scientists. This opinion is the result of a scrutiny of the roots of the model, the common interpretation of the phenomenon additivity, what it really tells us, and what is without sound basis.

There is a qualitative and a quantitative aspect of the phenomenon additivity. A *qualitative* analysis of combined effects investigates whether or not (yes or no) experimental data agree with a sham combination of one and the same substance (Loewe 1953, Berenbaum 1989). The *quantitative* approach considers additivity as the standard, with which deviations from the model can be compared quantitatively (e.g., Berenbaum 1989, Chou and Talalay 1984). This topic will be described and critically discussed in Chap. 3.

2.1.1 Historical perspective

The concept of additive combinations traces back to Frei (1913), at which time we did not have insight into binding of drugs to receptors and the knowledge about the agonist–agonist interaction. It was understood as a combined effect of A and B, where B was assumed to act just like A, i.e., like a dilution of A.

Let us first look at additivity from the fraction of *doses* of the parent compounds, yielding the same effect. For instance, if 1A and 1B alone produce 50% effect, 0.5A + 0.5B should again produce a 50% response, provided that A behaves like B. Also, other dose combinations, e.g., 0.2A + 0.8B should produce 50 % in this example. Isobolograms are based on this approach, described below (Fig. 4). We will here consider this procedure with respect to additivity.

2.1.2 The isobologram

Loewe and Muischnek (1926) reported on a graphical procedure to visualize combined effects by "isoboles" as lines which connect equieffective doses of A and B combined, expressed as fractions of equieffective doses of A and of B alone (Fig. 4). In principle, and on a phenomenological basis, a straight line isobole, i.e., a diagonal line between the equieffective doses of A and B alone (e.g., the $ED_{50}s$), was considered to indicate additive combinations. In isobolographic analysis, the terms ED_{50}, ED_{30}, etc. indicate the doses which in combination produce a fixed response. When DRCs of A are analyzed in the presence of an effective, fixed dose of B (e.g., Brunet et al. 1986), the $ED_{50}s$ of A + B do not represent half maximum effective doses of the respective DRC (see also Chap. 1.2.1). However, when fixed-ratio combinations (i.e., mixtures) are analyzed (e.g., Gessner 1988), there is no difference between a dose for the fixed response which is 50% and the half maximum effective dose, since the mixture curve shares E_{min} with the curves of A and B alone.

Isoboles, like curve 2 in Fig. 4, were considered to indicate synergism, curves 3 and 4 relative and absolute antagonism, respectively. Loewe and Muischnek (1926) invented the terms synergism and antagonism to describe the direction of deviations of isoboles from the additivity line. This circumstance can be seen as one source for the arising confusion and controversies in the scientific community between phenomenon and mechanism, since synergism and antagonism in isobolograms do not always correspond to enhancement and diminution of effects of one drug by another, described in Chap. 1. Furthermore, synergism and antagonism, defined as Loewe and Muischnek (1926) proposed, may be at variance with potentiation/synergism and antagonism in simple interaction, described and discussed in Chap. 1.4.1.

As a matter of fact, deviations from additivity in isobolograms indicate deviations from the sum of fractions of equieffective *doses*. Since the doses of A and B in combination are represented in isobolograms with which the same (magnitude of) effect can be seen, we can only assume that deviations in doses reflect greater or smaller effects in combination compared to those expected for additivity. Let us consider the fraction of doses of A and B which sum up to 1 as additive doses. If the same magnitude of effect is obtained with underadditive doses, the combined *effect* is regarded *overadditive*, since greater effects in combination than expected for additivity are assumed. In analogy, if the same effect is seen with overadditive doses, the combined *effect* is considered *underadditive*.

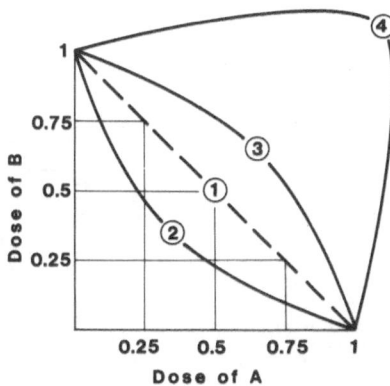

Fig. 4. Scheme of an isobologram with isoboles 1–4. Doses of A and of B = 1 are equieffective doses. The diagonal line (isobole 1) indicates a dose-additive combination, i.e., one achieved with additive doses. Isobole 2 illustrates an equieffective combination, which is achieved with lower than additive doses, whereas isoboles 3 and 4 indicate equieffective combinations for which higher than additive doses are necessary. Frequently, isoboles like isobole 2 are called overadditive, those like 3 and 4 underadditive, the latter were characterized as relative and absolute antagonism by Loewe and Muischnek (1926)

The isobologram technique is used by many investigators, mainly for evaluation of the effects of mixtures rather than the effect of a certain drug in the absence and presence of a fixed dose of another drug with similar effects. Where isoboles are derived from DRCs, which is often the case, the results of DRCs are reduced to a few points in isobolograms. It is also important to realize that isobolograms only consider additivity as the reference, with exceptions (see Pöch et al. 1990c). A critical evaluation of the isobologram approach is given in Chap. 9.

2.1.3 Dose-response curves

As already pointed out, the phenomenon of additivity can be explained by a simple model, in which substances A and B behave as if one of them is a dilution of the other one. Under this assumption, the combined effect of such substances must resemble a sham combination of one and the same compound. On this basis, the effect (E) of an additive combination can be calculated, e.g., for the combination of equieffective doses of A and B, termed 1 A and 1 B

$$E_{1A+1B} = E_{2A} \text{ and } E_{1A+1B} = E_{2B},$$

respectively. In an attempt to evaluate experimentally combined effects on the basis of DRCs, theoretical DRCs for additive combinations were constructed in order to compare observed DRCs of A in the presence of a fixed dose of B with expected additive DRCs (Pöch and Holzmann 1980/1981). A similar approach was followed by Draskoczy and Trendelenburg (1968) and by Kaumann et al. (1977) for special cases. Greater than additive com-

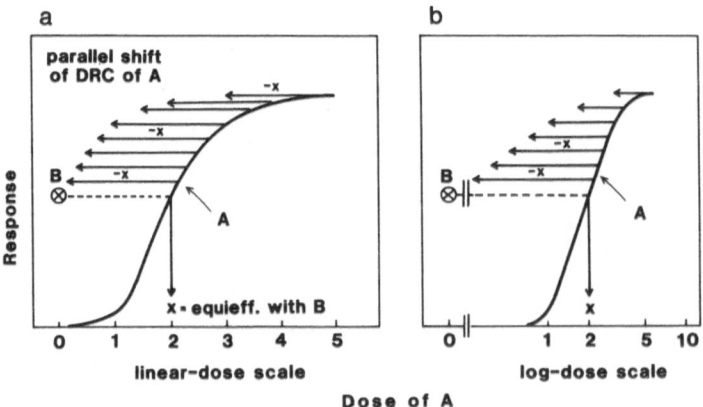

Fig. 5. Illustration of how dose-additive DRCs of A + B are obtained for DRCs of A with the doses on a linear (**a**) and on a log scale (**b**). Dose-additive DRCs of A in the presence of B (arrowheads) are shifted to the left by the dose x, representing the dose of A with which the fixed dose of B is equieffective. In **a**, this shift is in parallel with the DRC of A because x here at all doses of A represents the same distance – in contrast to **b**

binations are termed "overadditive" (supraadditive), weaker than additive combinations are termed "underadditive" (subadditive). In the analysis of DRCs the curves directly show effects above and below additive combinations, respectively.

During the last 10 years this methodological procedure was applied to many drugs in pharmacology as well as in other areas, where combined effects are investigated. The main part of this book rests on this principle, i.e., to evaluate combined effects by direct analysis of DRCs in comparison with additivity, representing "competitive synergism" (Ariëns et al. 1956a) as well as in comparison with independence (see Pöch 1991a). The term "competitive synergism" indicates the mechanism (competitive) as well as the phenomenon (synergism) of this interaction. It has been mathematically formulated by Ariëns et al. (1956a).

If a DRC of A is studied in the presence of a fixed dose of B, which behaves like A, and the dose of B equals a certain dose of A, x, the same effects are expected at doses of A minus x. This phenomenon is illustrated schematically in Fig. 5. It also illustrates the basis of a simple procedure to calculate and construct the theoretical additive DRC from the DRC of A alone and the effect of a fixed dose of B. Before continuing the description of additive DRCs, let us look at the underlying mechanism of additivity.

2.1.4 Mechanism underlying additivity

The additivity model describes the action in combination of agents which behave alike, just like the molecules of one and the same substance in a *sham combination.* The latter can be expected to behave in a dose-additive manner in any case, regardless of whether or not the molecules of a compound act reversibly or irreversibly, either bind to a single or to multiple sites in a specific or unspecific way. In any case, all the molecules of one and the same substance should behave alike. However, the question arises whether and when different compounds can mechanistically act together like a sham combination of one agent. It will be shown that there is convincing evidence for such an action only with compounds which reversibly interact at a single site.

Reversible action at a single site
When A as well as B reversibly bind to the same site, competitive interaction is expected which will result in an additive interaction if both agents are agonists. Since such binding requires certain molecular structures, it is a specific binding. Comparison of observed effects in combination with effects of additive combinations of agonists, therefore, allows a site-directed analysis (Table 1), in analogy to experiments with competitive antagonists. Briefly, effects which are significantly greater than additive, indicate that A and B differ in their molecular site of action. Underadditive effects do not necessarily point to different sites, since such combined effects occur with a full receptor agonist in the presence of a partial agonist, again by competitive interaction, described below.

Table 1. Overview of combined pharmacodynamic effects of A and B. Description, interpretation, and possible underlying mechanisms of interaction

Phenomenon	Interpretation	Mechanism
> Additive	not resembling A + A	different binding sites of A and B
Additive (dose-additive)	resembling A + A (sham combination, dilution)	same binding site of action[a] (competitive interaction)
< Additive	may resemble A + A in part (DRC parallel to additive DRC)	same binding site of action[a] (competitive interaction)
> Independent	enhanced net response ($ED_{50}\downarrow$, $E_{max}\uparrow$) *potentiation/synergism*	interaction[b]
Independent	unaffected net response to A, to B	no interaction[c] (different sites of action[c])
< Independent	diminished net response ($ED_{50}\uparrow$, $E_{max}\downarrow$) *antagonism*	interaction[c]

[a] If not independent
[b] In or between reaction pathway(s)
[c] If not additive (competitive)

Let us first consider agonist–agonist interactions, where the agonists, A and B, exhibit DRCs with the same steepness (slope) and the same maximum response (E_{max}). Hence, only a difference in the ED_{50} may occur, as in Fig. 6a. This is the case in which A and B convincingly behave alike, B in this example appears as a 5-fold dilution of A. The DRCs have to show the same slope, otherwise B could not be a certain dilution of A and could not resemble a true sham combination of A + A.

Then, if A is tested in the presence of B, the combined effects can be explained by competitive synergism (Ariëns et al. 1956a) and correspond to additive combinations (Fig. 6b). Hence, additive combinations of specific drugs point to the same binding site, provided the interaction is competitive in nature (Table 1). However, if an additive interaction at the same time corresponds to an independent action of A and B, the latter requires an action at different molecular sites. This phenomenon occurs with agents which exhibit steep DRCs (slope values around 1.6). In this case, neither a competitive interaction at the same site (additive interaction) nor an independent action at different sites can be assumed or excluded.

It is known in pharmacology that drugs which bind to the same receptor can differ in their intrinsic activity, i.e., in their E_{max}, as illustrated schematically in Fig. 6c. In this example, B acts as a full and A as a partial agonist. That means that binding of B to the receptor fully activates an effector system, whereas B only partially activates the same receptor-effector system. The combined effects of agonists with different abilities to elicit a biological response will depend on the proportions of these compounds present at the receptor which, in turn, depends on the testing conditions.

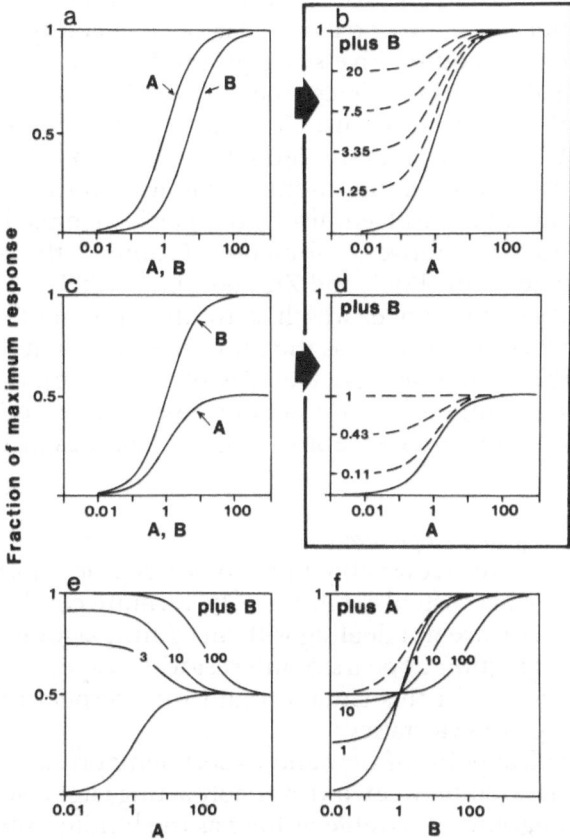

Fig. 6. Ideal DRCs of competitive agonist–agonist interactions (**b–f**). **a** The agonists A and B elicit the full maximum response = 1, and share the same slope = 1. **b** The interactions of A and B, whose DRCs are shown in **a**, expressed as effects of A in the absence (solid line) and presence of fixed doses (indicated by numbers) of B representing additive combinations of A plus B (– – –). **c** B is a full, A is a partial agonist, with shared slope = 1, the interactions of these agonists are shown in **d–f**. **d** Doses of the partial agonist A up to 1 B, equieffective with E_{max} of A, are additive, **e** whereas higher doses (3, 10, 100) of the full agonist B show effects which are progressively decreasing with increasing doses of the partial agonist A. **f** When tested in reversed order, the effects of B above B = 1 are antagonized by the partial agonist A, resulting in a parallel shift from the additive DRC (– – –), exemplified for the combination of B with 100 A

If the partial agonist A is tested in the presence of fixed doses of the full agonist B, the combined effect appears additive as long as the full agonist B is tested in doses which do not exhibit greater effects than E_{max} of A (Fig. 6d). On the other hand, if A is tested in the presence of doses of B, producing greater effects than E_{max} of A (Fig. 6e), competitive antagonism between A and B is seen, which corresponds to underadditive interactions when applied in the reversed order (Fig. 6f). The explanation is that B is displaced from the receptor by A with increasing doses of A, which at the most can produce a partial response, indicated by the lower E_{max} of A in Fig. 6c. Hence, the effects of a full and a partial agonist in combination reflect *competitive dualism,* i.e., competitive synergism in Fig. 6d and competitive antagonism in Fig. 6e. The latter is also seen in case the full agonist is tested in the presence of the partial agonist (Fig. 6f) at higher doses of B, i.e., at doses which singly produce greater effects than E_{max} of the partial agonist.

The competitive nature of this interaction is evident from the fact that the shift of the DRC of the full agonist by the partial agonist is in parallel with calculated DRCs of additive combinations (Table 1), the latter reflecting competitive synergism (Pöch and Zimmermann 1988).

In summary then, substances which share the same molecular site of action can interact competitively, as the molecules of A in a sham combination of reversibly acting agents. There is increasing pharmacological evidence that such competitive interactions also occur at other specific binding sites than transmitter receptors, e.g., at enzymes or ion channel proteins.

Irreversible binding to a single site

Interestingly, drugs can irreversibly bind to a receptor, as *antagonists* to which the respective agonists also bind, i.e., by forming covalent bonds via alkylation. So, here we are not dealing with an agonist–agonist interaction but this type of interaction appears of interest with respect to whether or not competition occurs in this situation and with respect to other irreversible actions, e.g., of cytostatics.

From a theoretical point of view and experimental results we can conclude that a competition between a (reversibly acting) agonist and an irreversibly acting antagonist is possible as long as the binding site is not irreversibly blocked (Ariëns et al. 1964b, Creveling et al. 1962). Further, experimental results show that the reversible inhibitor, butyrylcholine does protect the enzyme cholinesterase from irreversible inhibition of diisopropyl fluorophosphonate (DFP), however, not according to a strict competitive interaction (see Cohen et al. 1951: Fig. 1). Two irreversible receptor antagonists or enzyme inhibitors are likely to compete when present at the same time, as long as no irreversible reaction between the binding site and the attacking compound has taken place.

In summary, competitive interactions between reversibly and irreversibly acting agents, and between irreversible blockers can occur only under certain circumstances. Hence, additive interactions will here reflect competitive interaction in certain situations, only.

Reversible interactions at multiple sites

Present day knowledge suggests that highly lipophilic agents, like anesthetics, *nonselectively* but reversibly bind to multiple hydrophobic sites (Rubin et al. 1991). Namely, it has been observed recently that "unspecific" drugs and ethanol bind to *various* "receptive sites" on ion-channels (Dilger and Firestone 1990). This implies that so-called nonspecific drugs and other lipophilic chemical compounds can bind to different hydrophobic sites or domains on various ion channels, e.g., located on the GABA- and the NMDA-channel (Gonzales and Hoffman 1991) at the same time. Hence, the action of these compounds is better described as nonselective.

The close correlation between the physical property of lipophilicity and effect can thus be explained in the following way. With increasing lipophilicity, the molecules have a greater chance to reach the hydrophobic sites to which they eventually bind. Hence, lipophilicity is required but in itself does not explain the action of compounds which have been termed "unspecific" in textbooks (e.g., Gero 1971).

Now, if we consider that the binding of anesthetics and other highly lipid soluble agents to hydrophobic sites on ion-channels (and other proteins) is rather *nonselective*, then we can reasonably speculate how they can interact. They are expected not to interact in general at one and the same binding site, as specific drugs often do, but to bind to different sites, either on different proteins or at distinct sites located on the same (ion-channel) protein. Then, they are not expected to interact competitively but could interact by allosteric mechanisms by binding to nearby sites.

We have recently obtained evidence that overadditive combined effects can occur when drugs are tested in combination which share the same action but where one drug possesses an additional action (Holzmann et al. 1992). Interestingly, blockade of this additional action resulted in additive combinations (Holzmann et al. 1992). So, we cannot, from a mechanistic point of view, expect competitive interactions (resulting in additive combinations) to occur with unspecific agents. Hence, combinations which appear additive are then expected to occur by mere coincidence.

Experimental evidence, phenomenologically, showed *additive combinations* between "unspecific" drugs. As an example, the interaction of general anesthetics was reported to be additive for many combinations, like ether plus chloroform in contrast to overadditive combinations of nonspecific plus specific drugs, like urethane plus morphine (Bürgi 1938). The clinical consensus is that general anesthetics exhibit additive combinations. However, deviations from a simple additive behavior have been reported (Cole et al. 1989, 1990) and discussed as possibly caused by differences in pharmacokinetics and/or by additional effects (see Cole et al. 1989, 1990).

Evidence for non-selective binding of lipophilic drugs comes from recent DRC-experiments in our laboratory with organic solvent-induced hemolysis of human erythrocytes. Effects which clearly deviate from (dose-)additive were observed with certain combinations (Köck unpubl. results).

A possible explanation for combinations of unspecific agents to appear additive is that the combined effects somewhat exceed independent effects, as is often seen (Pöch 1991 b) and that theoretical effects of additive combinations also exceed independent effects. The latter presumption requires DRCs with greater steepness than 1.6, and this has been observed experimentally (Rang 1960).

With "unspecific" compounds greater steepness of DRCs than with receptor agonists binding to a single site has been observed. For instance, from the DRCs published by Rang (1960), it can be deduced that the slope of DRCs for all effects studied was above 1.9, in the range between 1.9 to 3.6 for smooth muscle relaxant effects, and around 3.4 for inhibition of oxygen consumption. This fact may also point to more than one single site of action. Let us take a mixture of two or more substances with different actions as if it were one substance with different actions. Now, let us assume that A and B show the same steepness. In combination of equi-effective doses we can expect a greater steepness of the DRC, i.e., of the mixture A + B, if they interact independently. Namely, at the threshold dose of the components, no greater effect in combination is expected but at the half maximum effective dose, the calculated effect of the mixture is 75%.

Irreversible interactions at multiple sites

Since combinations of reversibly acting drugs at multiple sites and the combined effects of irreversibly acting agents at a single site do not (truely) reflect a sham combination, compounds acting irreversibly at multiple sites are even less likely to resemble a sham combination mechanistically. There is experimental evidence that many cytostatics irreversibly attack multiple sites, i.e., alkylate different atoms of the bases of the DNA to a different degree (Singer 1982, cited in Birgersson et al. 1988).

2.1.5 Additive dose-response curves

Figure 7 shows families of response curves which represent competitive synergism, i.e., dose-additive combinations. These DRCs correspond to sham combinations which presumes that B shares the same steepness with A of DRCs. Competitive synergism is exemplified for effects of B of 1, 10, 20, 40, 60, and 80% of the maximum obtained with A. It can be seen that all DRCs reach the same maximum as A alone. On the left side, this maximum equals the possible maximum; on the right side, it equals 66% of the possible maximum, chosen in an arbitrary manner for demonstration purposes.

Furthermore, the dependence of additive DRCs on the steepness of the DRCs of A is exemplified for a slope of 0.8 (Fig. 7a and b), for 1.6 (Fig. 7c and d), and for 5 (Fig. 7e and f). These control DRCs in slope represent flat, steep and very steep DRCs, frequently observed with drugs in pharmacology (a and b), with cytostatics (c and d), and with toxic agents (e and f). Interestingly, additive DRCs do not show the same steepness as the DRCs of A alone, except at a slope of 1. With increasing effects of B alone the slopes tend toward a slope of 1.

Also, differences in the ED_{50} of the additive DRCs compared to the ED_{50} of A alone are evident, which are marked for flat and very steep DRCs, respectively (Fig. 7). In flat DRCs, the ED_{50}s of additive DRCs are progressively shifted to higher concentrations of A (Fig. 7a and b), whereas in very steep DRCs the shift of the additive ED_{50}s occurs in the opposite direction, i.e., to lower concentrations (Fig. 7e and f). There is not much change in these values in case of DRCs with a slope of 1.6 (Fig. 7c and d).

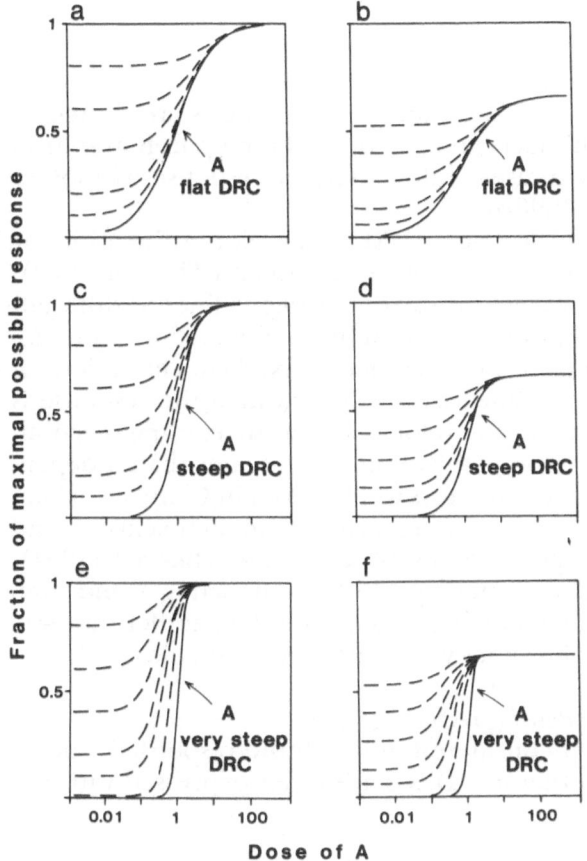

Fig. 7. Additive DRCs of A in the presence of B (broken lines) for flat (**a, b**), steep (**c, d**), and very steep DRCs of A (**e, f**) for effects of B corresponding to 10, 20, 40, 60, and 80% (**a–d**), and for 1% in addition (**e, f**) of E_{max} of A. Note that all additive DRCs reach the same maximum as A alone, irrespective of the height of E_{max}, which equals 100% (**a, c, e**), and 66% of the possible maximum (**b, d, f**), given as an example

In summary, all additive DRCs reach the same maximum as the DRCs of A but, depending on the steepness of the DRCs, changes in slope and $ED_{50}s$ may occur. A systematic investigation revealed changes in slope and ED_{50}-values as a function of the slope of A and the effect of B. Tabulated values for the respective additive DRCs are shown in Appendix B, and allow a simplified construction of additive DRCs by ALLFIT compared to the one already published (Pöch et al. 1990a).

Additive DRCs at threshold doses of B
Another interesting aspect of dose-additive combinations is the fact that in case of very steep DRCs of A, calculated additive DRCs are more or less shifted to the left from the DRC of A alone at threshold doses of B (Fig. 7e and f). This peculiar phenomenon is explained by the fact that in very steep DRCs, a change in doses produces marked changes in response; this is discussed further in Chap. 3.3.2.

So, there is kind of overlapping between simple and complex interaction in that sense that we have to consider an additive response despite the fact that B only produces a threshold effect instead of no effect in simple interactions.

Additive DRCs with linear-dose scale
As shown in Fig. 5a, DRCs of dose-additive combinations are shifted in parallel to the DRC of A. This fact can be exploited for a purely graphical estimation of additive DRCs in case the doses and the response are plotted on linear scales (Pöch et al. 1990b).

A special form of a linear dose-scale DRC is the plot of the inhibitor concentration vs. $1/v$ in enzyme reactions (Yonetani and Theorell 1964), resembling a Dixon plot. However, instead of straight lines at different *substrate* concentrations, as in the Dixon plot, lines obtained with one inhibitor are shown in the absence and presence of fixed concentrations of another *inhibitor*. This plot is a linearized DRC in which the response is expressed by reciprocal values of the velocity (v) of the enzyme reaction. An example for an additive (competitive) and for greater than independent interaction of two enzyme inhibitors will be given in Chap. 8.1. Competitive and independent interactions were termed "mutually exclusive" and "mutually nonexclusive", respectively, by Yonetani and Theorell (1964), followed by Wong (1975). The former type, with agents acting at and competing for the same molecular site, is characterized by parallel "dose-response" lines, the latter exhibits lines which intersect at the dose axis.

Additive DRCs with linear-dose log-response graphs
So far, only DRCs have been considered in which the response is shown on a linear scale. In chemotherapy, DRCs are often expressed with the

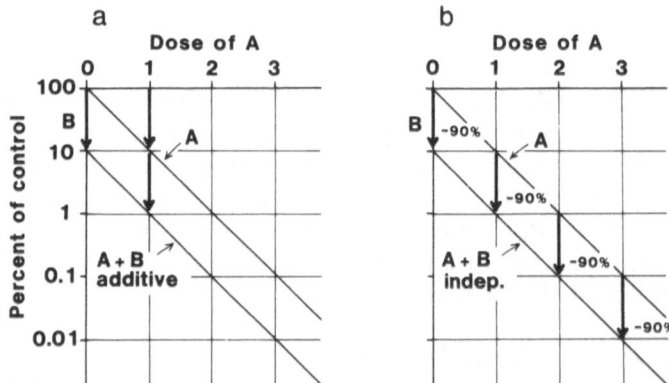

Fig. 8. Linear-dose log-response curves, used to demonstrate effects of cytostatics and other chemotherapeutics, for the doses of 1, 2, and 3 of A (x-axis) and the response expressed as percent of control (y-axis), e.g., cell survival. The dose of 1A reduces the control response of 100% to 10%, i.e., by 90%, the doses of 2A, and 3A exhibit the same relative effect, resulting in an exponential, linear decrease of the control in **a** as well as in **b**. The lower dose-response line shows the effect of A + B, at a dose of B which is equieffective with 1A, corresponding to an additive (**a**) as well as to an independent combination (**b**)

doses on a linear and the response on a log scale (Fig. 8). Often, a straight line of dose response is seen in such a graph with cytostatic agents, which represents an exponential DRC. Then the dose-additive curve of A in the presence of B is a straight line also, and it is in parallel with the line for a drug A, originating at the response to B alone (Fig. 8a). The combined effect of additive combinations can there be seen as the addition of "effects". Therefore, additivity in this graph corresponds to the "effect summation" model.

2.1.6 Comparison of effect-additive with dose-additive combinations

Quite often, the term additive is used to describe combined effects which are equal to the addition of effects on a phenomenological basis. However, conclusions with respect to mechanisms are not infrequently drawn from effect-addition, or from combinations corresponding to or exceeding the sum of individual effects. Hence, it appears appropriate to compare dose-additive and effect-additive combinations. The phenomenon of effect-addition *can* be observed with linear-dose log-response graphs, mentioned before, in additive interactions (Fig. 8b).

With nonexponential curves, and if the effects of compounds or factors are very small with respect to the possible maximum, we can see that independent interactions can be approximately described by the sum of the individual effects (see Chap. 2.3.1).

When inhibitory effects are expressed on a log scale, described above, equal changes reflect equal effects (Fig. 8). If, for example, A and B are equieffectively inhibitory and B reduces the control response from 100 to 10, then by an independent action A will lead to a reduction from 10 to 1. This means that A plus B together reduce the control value from 100 to 1, which on a log scale is the sum of the reduction from 100 to 10 (produced by A) and 100 to 10 (produced by B) (Fig. 8). If, in addition, the dose response of A follows an exponential curve, i.e., a straight line in such a graph, then the sum of effects can also be explained by a dose-additive interaction (Fig. 8a). However, this is not the case with non-exponential curves of a slope of considerably less or more than 1.6, often observed in other fields than chemotherapy.

Table 2. Relationship of DRC steepness and dose-additive interactions

E_A, E_B (%)	Slope	E_{A+B} (%)	
		dose-additive	effect-additive
25	0.8	37	50
25	1.0	40	50
25	3.0	73	50
25	5.0	91	50

Depending on the steepness of DRCs dose-additive interactions can produce definitely smaller or greater effects than the sum of individual effects. This situation is exemplified in Table 2 for a slope between 0.8 and 5, tested in the presence of B for the combined effect of A and B, which separately produce 25%. Whereas effect-additive combinations yield 50% (25 + 25%), dose-additive combinations produce 37 to 91% in this example. Thus, it is clear that the magnitude of dose-additive combinations may be quite different from the one expected for effect-addition. DRCs with slopes around or below 1 are quite frequently observed in pharmacology, notably with agents acting at transmitter receptors. Slopes of 3 to 5 or above are quite frequently seen with agents producing toxic actions. So the examples listed here represent real DRCs.

2.2 Site-directed analysis

In view of the importance in evaluating the site of action of drugs and other chemicals, a brief outline for a site-directed analysis is presented here which includes testing for potentiation in simple interactions as well experiments with antagonists. Practical examples are presented in Chap. 8.

Simple interactions

1) If B interacts with A as a competitive antagonist, A must bind to the same site as B.
2) If B interacts with A as a noncompetitive antagonist, the same site of interaction is excluded.
3) Likewise, if B shifts the DRC of A to the left, the same site of action is excluded.

Complex interactions

1) If A in the presence of B follows an additive interaction, the same site of action is possible, but not proven.
2) If A in the presence of B exhibits an underadditive interaction, and B possesses partial agonist acitivity, the same site of action can be proven by demonstration of competitive interaction.
3) If the effect of A in the presence of B is significantly overadditive, the same site of action is excluded – unless this finding can be explained by differences in pharmacokinetics.

There is still another problem in the evaluation of additive combinations, touched on in Chap. 2.1.4. Additive combinations may correspond to an independent interaction as well, which is quite frequently seen with chemotherapeutic agents. In this situation, no conclusions as to the site of action can be drawn. In other words, A and B may or may not bind to the same site. The relationship between additivity and independence will be described and discussed below (Chap. 2.4).

2.3 Concept of independence

Another basic concept of combined effects, that can be expressed mathematically, is that of independent interaction between drugs, chemicals, and physical factors (Bliss 1939). It can be regarded as a simple model for those compounds or physical factors which produce similar effects not by acting at the same site but at different sites, with the notable exception of compounds exhibiting exponential DRCs which could as well act at the same site, resulting in the phenomenon of independent interaction (Berenbaum 1981).

Ariëns et al. (1956b) used the term "functional synergism". This terminology was followed by Pöch and Holzmann (1980/1981). We now prefer the terms "independent action", "independent interaction", or "independent combination", and avoid the term "synergism" in this context.

We can surmise that two drugs acting at completely different sites could in fact act independently of each other when combined. Although we *cannot* expect them in general to act independently when administered together to produce a common effect, a comparison with independent combination nevertheless seems worthwile, since independent effects represent an unaffected response, as will be explained below.

The effects of A may not be affected by B, when B by itself does not exhibit an effect. There, the action of A is clearly independent of B. If B by itself causes an effect in the same or in the opposite direction (small arrows in Fig. 9), A may still act independently of B by activating a different and independent pathway. However, we cannot expect that the "absolute" effects of A are unaltered if B alone causes an effect in the same or in the

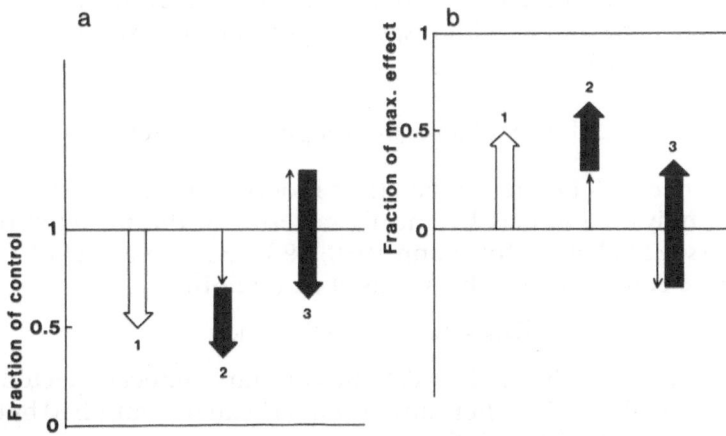

Fig. 9. Schematic illustration of unaltered response to A (arrow *1*) in the presence of B (arrow *2, 3*) from decreased or elevated baseline, induced by B (thin arrows): **a** inhibition, expressed by the fraction of control **b** activation, expressed as fraction of the maximum possible effect. The response to A in the presence of B (*2, 3*) is characterized by a decrease to 0.5 of the control (**a**) and an increase to 0.5 of the maximum (**b**). The same response to A is seen in the absence of B (*1*)

opposite direction, respectively (Fig. 9). The large black arrows in Fig. 9
indicate an unaltered action of A in the presence of B, as will be explained
below. The changes themselves, however, are smaller in the presence of B
acting in the same direction but larger in the presence of B acting in the
opposite direction. This phenomenon is quite frequently observed; it led
Wilder to formulate a "law of initial value" (Wilder 1967). However, as
might be expected, there are also a number of experimental results in
disagreement with this "law" (Pöch and Brunner 1984).

Let us now explain why the effects of A in the presence of B in Fig. 9 are
mechanistically independent from B.

The concept of independent interaction requires that the relative effects, i.e., the net effects of A are not altered by B. This is true for different
situations, e.g., for the case in which the response is expressed as *fraction of
control* (Fig. 9a) or as *fraction of the maximum possible effect* (Fig. 9b), irrespective of the direction of effects of B (thin arrows). Arrows 1 to 3 show that A
in the absence as well as in the presence of B exhibits a 50% reduction from
the initial control value (arrow 1) as well as from a decreased and increased
"control" value (arrows 2 and 3), respectively. The same relative effects of
A in the presence of B are also seen in Fig. 9b. Hence, irrespective of
whether the response to A is inhibition (Fig. 9a) or activation (Fig. 9b), the
relative effects of A are unaltered by B with respect to the possible minimum (Fig. 9a) or maximum (Fig. 9b). Also, the net effects of B, again with
respect to the maximum, are not altered. The concept shown in Fig. 9 is
schematically illustrated in Fig. 10 where the responses to B are given by the
arrowheads.

Since independent effects represent unaffected actions in combination, they also represent *no interaction*. Hence, independence is the appropriate reference for a quantitative evaluation with respect to enhanced
or diminished effects, i.e., with respect to potentiation/synergism and antagonism (Chap. 3).

2.3.1 Calculation of independent effects

The combined independent total effect E of A and B (E_{A+B}) is calculated
from the individual effects E_A and E_B in terms of the maximal possible
effect = 1 (see Pöch and Holzmann 1980/1981, Pöch et al. 1990b). This is
basically the equation given by Ariëns et al. (1956b):

$$E_{A+B} = E_A + E_B - (E_A \times E_B). \tag{1}$$

For instance, if $E_A = 0.5$, and $E_B = 0.4$, the combined effect expected equals
0.7 ($E_{A+B} = 0.5 + 0.4 - 0.2$). Alternatively, Eq. (1) can be simplified by expression of values of $(1 - E)$, as follows:

$$(1 - E)_{A+B} = (1 - E)_A \times (1 - E)_B. \tag{2}$$

Equation (1) convincingly shows that if the individual values are very small
with respect to the maximum, the combined effects can roughly be described by the summation of effects:

$$E_{A+B} = E_A + E_B. \tag{3}$$

This fact is illustrated by the following example, in which the individual effects are each assumed to produce only 2%. $E_{A+B} = 0.02 + 0.02 = 0.04$, according to Eq. (3) is practically identical to E_{A+B} equals $0.02 + 0.02 - (0.0004) = 0.0396$, obtained by Eq. (1).

Calculation of combined reduction of control values (inhibition) by A and B can also be calculated simply, and correctly, by

$$FC_{A+B} = FC_A \times FC_B \tag{4}$$

where FC_{A+B}, FC_A, and FC_B denote fractions of the control value C (at C = 1). For instance, if A as well as B reduce the control value to 0.5, FC_{A+B} equals 0.25 for an independent interaction.

It should be noted here that Eq. (4) for independent actions in combination can be described by the "multiplicative model". If the effects rather than the fraction-of-control values are used to calculate independent effects, we have to use Eq. (1), where no multiplication of effects can be seen. In special situations, Eq. (3) may be applicable, which describes the summation of effects, i.e., an effect-additive combination.

On the basis of the multiplicative Eqs. (2) and (4), simple calculation of independent interactions with more than two factors is possible, as exemplified for 3 factors, A, B, and C in Eqs. (5) and (6).

$$(1 - E)_{A+B+C} = (1 - E)_A \times (1 - E)_B \times (1 - E)_C, \tag{5}$$

$$FC_{A+B+C} = FC_A \times FC_B \times FC_C. \tag{6}$$

Independent effects corrected for "background"
In case of a considerable "background" (x) the effects E can and should be corrected, e.g., when Eqs. (1), (2), or (5) are applied:

$$\text{Corrected } E = (E - x) / (1 - x).$$

Example: "background" = 0.2. E_A, $E_B = 0.5$, corrected E_A, $E_B = 0.3/0.8 = 0.375$, corrected independent effect = 0.61 (calculated by Eq. (1) with corrected values of E_A and E_B; the respective uncorrected independent combination is $0.5 + 0.5 - 0.25 = 0.75$).

There are probably many situations in which corrections are indicated, e.g., in clinical pharmacology with marked placebo effect as "background" or in risk evaluation with basal risk as "background".

2.3.2 Independent dose-response curves

The DRCs of an independent action of A and B can rather simply be constructed under the supposition that we can express the response with respect to the possible maximum, e.g., as in Fig. 10. As mentioned, we will use the term "independent DRCs" for DRCs of independently interacting agents or factors.

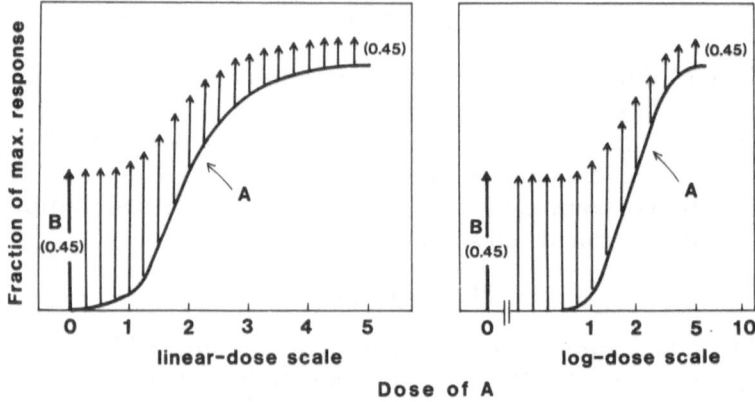

Fig. 10. Illustration of how independent DRCs of A in the presence of B are obtained, exemplified for B causing 45% of maximum increase in the absence and presence of various doses of A, analogously presented as in Fig. 5. Since B acts independently of A, the fraction of maximal response to B is 0.45 from 0 dose of A to the highest dose of A. Linear- and log-dose scales are given to allow immediate comparison with Fig. 5

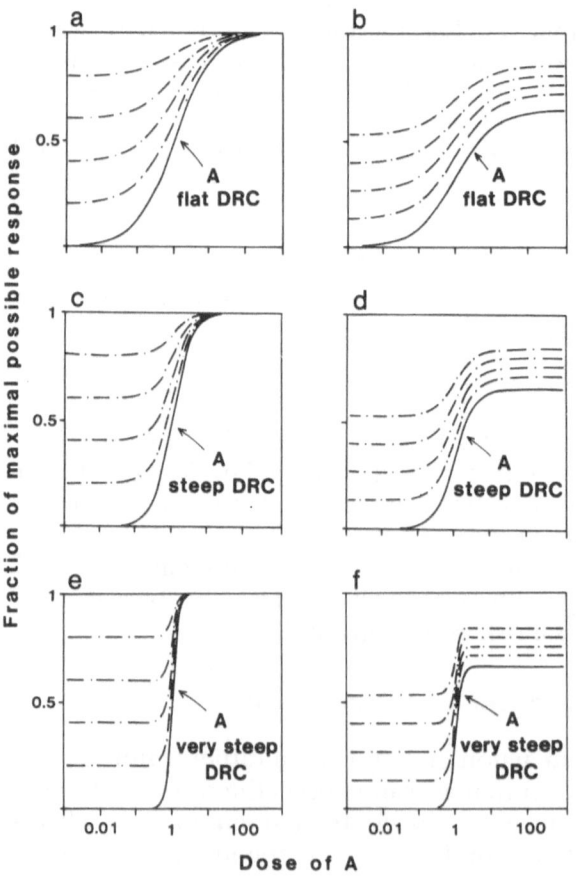

Fig. 11. Independent DRCs for the same examples of DRCs of A as in Fig. 7 (broken dotted lines). All independent DRCs share the same ED_{50} and the same slope with the respective curves of A alone. **a, c, e** A alone has reached 100%, therefore, no further increase by B is possible. **b, d, f** A alone reaches 66%, therefore, E_{max} of independent combinations is increased by the same fraction of maximum possible increase as exhibited by B alone (see Fig. 10)

In Fig. 10 the construction of independent DRCs is schematically shown for a drug A in the presence of a fixed dose of B. The arrowheads indicate points of the respective independent DRC of the combination of A and B. Such points of an independent interaction can easily be calculated by various equations. As a matter of fact, numerous equations can be found in the literature for calculation of independent effects, of which examples were given above for substances or factors, A, B, and C, respectively.

To allow a direct comparison with additive DRCs, independent DRCs in Fig. 11 are analogously shown for the same DRCs of A as in Fig. 7. In contrast to the additive DRCs, all DRCs of independent interaction exhibit the same steepness and the same ED_{50} ($[E_{min} + E_{max}] / 2$) as the DRCs of A (Fig. 11). Independent DRCs can in all situations be seen as an upward shift of the DRCs of A alone by the fraction of increase produced by B alone (Fig. 10).

Differences between theoretical DRCs for independent and additive combinations are especially pronounced in situations in which A alone does not reach the maximum possible effect (Figs. 7 and 11, panels b, d, and f). The comparison of independent and dose-additive combinations is the topic of the following section.

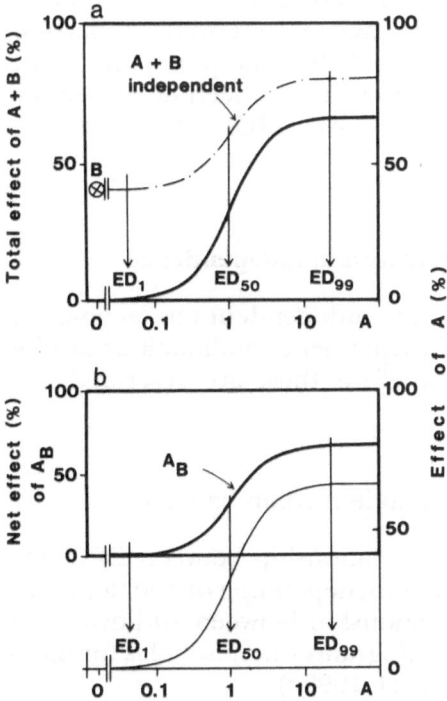

Fig. 12. Unaffected DRCs of A in the presence of B, expressed as total effects of A + B (**a**) or as net effects A_B (**b**). **a** The total effects of A + B have been constructed under the assumption of independent interaction between A and B. ⊗ Effect of B alone; solid line, A alone. **b** The net effects show the corresponding relative effects of A in the presence of B (A_B) (left scale) which equal the effects of A alone (right scale), repeated from **a**

As pointed out already, the relative effects of A in independent combinations appear unaltered by B, although an increase in E_{max} of total effects occurs where this is possible. This important fact is schematically illustrated in Fig. 12. With respect to the doses, we see that not only the ED_{50} ($[E_{min} + E_{max}]$ / 2) is unaltered in independent interaction, but that the ED_1 and the ED_{99} is neither increased nor decreased. All other ED-values are also unchanged by an independently acting agent (not shown).

From the foregoing considerations it becomes clear that, no matter which effect B alone produces, the DRCs of *net effects* of A in the presence of B match with the DRC of A alone. Hence, all the characteristics of A are unchanged by B, i.e., E_{min}, E_{max}, ED_{50}, and slope, indicated by A_B in Fig. 12. Imagine the effect scale for A_B to be extended to the effect scale of A in Fig. 12 b, then the curve of A_B matches with the curve of A alone. Hence, the net effects A_B, i.e., the net effects of A in the presence of B, equal the effects of A in the absence of B.

Independent DRCs with linear-dose log-response curves
When the *response* is plotted on a log scale (Fig. 8), the same graphical reduction of response represents the same effect. For instance, a reduction from 100 to 10 equals a reduction from 10 to 1, and from 1 to 0.1. Hence, if B causes a reduction by 90% from 100 to 10, then B in the presence of A shows the same reduction by 90% from 10 to 1, and from 1 to 0.1, as illustrated in Fig. 8 b. Hence, the independent DRC of A + B is parallel to the DRC of A (Fig. 8 b) – and indistinguishable from the respective additive DRC (Fig. 8 a). As a matter of fact, there is no difference between dose-additive and independent curves of A and B, as is evident from comparison of Fig. 8a and b, if A exhibits an *exponential* DRC. The phenomenon of coinciding DRCs of independence and additivity has been described for effects expressed on a linear scale with nonlinear DRCs (Unkelbach and Pöch 1988).

2.4 Relationship between additivity and independence

The relationship between dose-additive and independent interactions on the one hand and effect-additive and independent combinations on the other hand is completely different. Therefore, these are described and discussed separately.

2.4.1 Dose-additive and independent combinations

Figure 13 shows that there is no simple relationship between DRCs for dose-additive and independent combinations, depending on the slope and on E_{max} of A. This slope-dependent relationship between additivity and independence can also be shown in isobolograms when isoboles for independent actions are constructed (Pöch et al. 1990 c).

To illustrate the varying relationship between these two phenomena, DRCs are shown here, rather than isoboles. Two pairs of additive and independent DRCs of Figs. 7 and 11 have been superimposed in Fig. 13. Where the maximum possible effect is already reached with A alone, the two theoretical curves are almost identical with drugs exhibiting an exponential (-type) DRC, i.e., with a slope of 1.6 (Fig. 13c). In all other situations shown in Fig. 13, more or less marked differences between the two theoretical curves are evident. It is interesting that independent effects are greater than expected for dose-additive interactions (Fig. 13a and b) where A exhibits a "flat" DRC. In case of "steep" DRCs, the differences depend on E_{max} of A. Only with lower than possible E_{max} will *independent effects* exceed those of additive combinations (Fig. 13d). In case of "very steep" DRCs, *additive* DRCs exceed independent DRCs, at least in the submaximum effective dose range of A (Fig. 13e and f).

This comparison indicates that we can expect that in some cases independent interactions correspond to overadditive combinations (Fig. 13a and b) or that there may be hardly any difference between them (Fig. 13c)

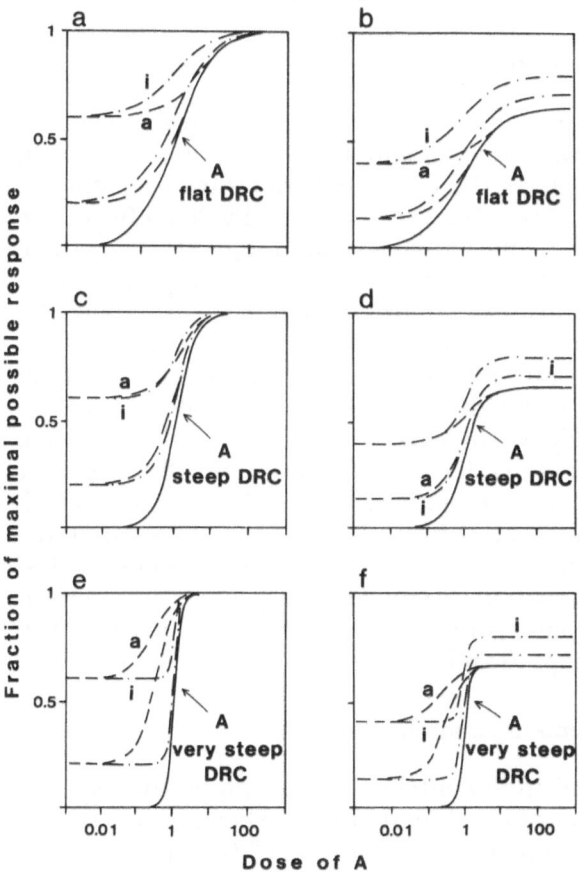

Fig. 13. Relationship between additive and independent DRCs exemplified for effects of B alone amounting to 20 and 60% of E_{max} of A, analogous to Figs. 7 and 11. **a, c, e** A alone reaches the possible maximum, **b, d, f** E_{max} of A alone corresponds to 0.66 of the possible maximum. *a* Additive curves, *i* independent curves

Table 3. Comparison of effect-additive and independent effects with respect to individual effects of A and B

E_A (%)	E_B (%)	E_{A+B}	
		effect-additive	independent
10	10	20	19
20	20	40	36
30	30	60	51
40	40	80	64
50	50	100	75

or that additive interactions represent greater effects than independent combinations (Fig. 13e). These findings have great theoretical and practical implications for a proper qualitative and quantitative analysis of combined effects. For instance, observed combined effects of A and B may at the same time appear additive and independent. However, this is seen only with compounds exhibiting steep DRCs with slope values around 1.6 *and* with the same E_{max}, e.g., in viability studies with cytostatics (Berenbaum 1981).

2.4.2 Effect-additive and independent combinations

There is a much simpler relationship between the sum of individual effects, i.e., the addition of effects, and independent actions. As a matter of fact, there is not much difference when the individual effects (or frequency of effects) are very low (see Eq. (3)). The greater the individual effects, the greater the differences between effect-additive and independent combinations, since independent effects are related to the possible maximum in response (see Eq. (1)). To illustrate the differences, let us compare the response, expressed as percent of maximum, of equieffective doses of A and B in combination. A comparison with greater individual effect values than 50% is not appropriate when 100% is the maximum possible effect (Table 3).

From a practical point of view, important conclusions can be derived. (1) Combined effects greater than effect-additive represent effects greater than expected for independent interactions. This is also true for experiments in which no maximum in response can be taken into account, described and discussed in Chap. 3.3.2. (2) Combined effects which are less than effect-additive do not allow general conclusions as they may represent combined effects greater, equal to, or smaller than independent. Practical examples will be given in Chaps. 6 and 8.

2.5 Summary and conclusions

The models of additivity and independence are of importance for the evaluation and interpretation of combined effects. They are based on a "dose approach" (additivity) and an "effect approach" (independence).

In the opinion of the author, the *additivity* concept should only be applied where one is interested whether the agents under study share the same *site of action*, more specifically, a single site of action (Table 1). The latter requires that the individual DRCs show the same slope. Otherwise, the presumption that one agent can be regarded as a dilution of the other is not met. If observed effects significantly deviate from additivity, different sites for A and B can be concluded safely, whereas experimental combinations which appear additive, may or may not be caused by competition at a single site. These conclusions require that no changes in concentrations of A and B occur by pharmacokinetic interactions.

If we are primarily interested in the *doses* in combination for a given effect, and in a suitable way of expression, we can use the sum of fractions of equieffective doses, the FEC-index, mainly used as FIC-index (sum of fractions of equieffective inhibitory concentrations) in chemotherapy (Elion et al. 1954, Hall et al. 1983). Also, the equivalent combination index (CI) (see Chap. 3.6.1) could be used as descriptive term.

A comparison with *independent effects* appears of value where we are interested whether or not the *effects* of one agent are affected, i.e., enhanced or diminished by another agent (Table 1). The model of independence does not require that a chemical compound binds to one or more sites on a protein, hence it can be applied to study interactions of physical or risk factors as well (Pöch 1991 b). When agents or factors in combination produce greater than independent effects, it means greater net effects in combination than in the absence of other agents or factors. Since greater net effects in dose-response studies by necessity are associated with reduced ED_{50} and/or enhanced E_{max} values, potentiation is evident in this situation, as will be described and discussed in the following chapter.

2.6 Tips and hints

It appears that many researchers prefer one of the two models, additivity or independence, on the basis of the idea that one model is better than the other. It has been shown here that a comparison of experimental combined effects with the models tells us different things as we either take the effects of doses of a sham combination or the unchanged relative effects for comparison. Hence, we have to consider what the models tell us, not which model is better. For instance, a deviation from a sham combination tells us that A and B do not act alike, excluding an action at the same molecular site, greater or smaller relative effects in combination indicate enhanced or diminished effects.

If we are interested in the site of action, a comparison with additivity is appropriate, whereas independence is the appropriate reference whenever we are interested in enhancement or diminution, i.e., whether A and B interact in a synergistic or antagonistic manner. Many investigators compare experimental combined effects with additivity in its broader meaning, irrespective of whether or not the same site of molecular action is excluded. Here, additivity has to be considered as a mere phenomenon with many possible underlying mechanisms of action.

A survey of the literature suggests that most publications deal with or are based on the additivity concept rather than on the model of independent action or other types of interactions. For readers who are interested in those methodological papers, relevant publications are listed here in alphabetical order: Ashford and Smith (1964), Berenbaum (1977, 1978, 1980, 1985, 1988, 1989), Bödeker et al. (1990), Carter (1985), Carter and Carchman (1988), Carter et al. (1988), Chou and Talalay (1977, 1981, 1983, 1984), Greco (1987), Greco and Lawrence (1988), Hewlett and Plackett (1950, 1956, 1979), Hoel (1987), Könemann (1981), Kundi (1987), Leff (1987), Loewe (1953), Loewe and Muischnek (1926), Ohashi (1976), Plackett and Hewlett (1952), Steel (1979), Steel and Peckham (1979), Sühnel (1990, 1992), Tallarida et al. (1989), Undem and Adams (1988), Unkelbach and Wolf (1984, 1985a), Van den Brink (1973), Ward (1988).

3 Synergism/potentiation and antagonism – phenomena and mechanisms

This chapter describes and discusses the differentiation and quantitation of combined effects with respect to synergism or potentiation on the one hand, and antagonism on the other, in *complex interactions*. In the combinations considered here, the agents exhibit effects in the same direction, when tested singly. Hence, in binary mixtures, both compounds are active. This is the situation which most researchers associate with the expression of "combined effects".

Colleagues in this field are aware of the different use and definitions of the terms synergism or potentiation and antagonism and of the disagreement among investigators. One and the same result can indeed be interpreted differently (Berenbaum 1989). To understand this problem we will start this chapter with a description and discussion of the present unsatisfactory situation for differentiation and quantitative expression of combined effects. It may help the reader to note whether phenomena or mechanisms are described and interpreted, respectively. Also, there are two approaches to this topic, one by consideration of changes in effects at given doses, the other by looking for changes in doses at a given effect. For convenience, the first approach is termed "effect approach", the second one "dose approach".

3.1 The current dilemma

Chapter 1 already dealt with the terms synergism/potentiation and antagonism in *simple interactions* (Chap. 1.4). There, examples for types of potentiation were described on the basis of a decrease in the ED_{50} and/or an increase in E_{max} of a DRC of A, caused by B. This characterization of potentiation is generally accepted, with some discussions about the preferred term for that phenomenon. It is based on the enhancement of effects of A by B (potentiation) and on the mirrorlike diminution of effects (antagonism).

Complex interactions are evaluated and interpreted differently by different researchers, as mentioned. For instance, many rely on additivity, others on independence as standard. This is one aspect of the current dilemma. Another aspect is the different characterization by many researchers of synergism/potentiation and antagonism in complex compared with simple interactions. We might suspect that one problem is linked to the other one, and we may ask whether there is a way out of this dilemma. It appears that the problem could be overcome on the basis of evaluation of *effects*, the appraisal of which is described below.

3.1.1 The "effect approach"

Sum of individual effects
Some investigators compare combined effects with the sum of individual effects, as described in Chaps. 2.1.6 and 2.4.2. Although this approach may appear simpleminded, it is not the worst one, at all. It is not based on a mechanistic model, but simply compares descriptively the combined effects with the calculated sum of individual effects. In case that combined effects exceed the sum of individual effects, it indirectly gives us the information that the combined effects must also exceed the expected effects of independently acting compounds (Chap. 2.4.2).

Net effects
There are others who characterize potentiation as net effects of A in combination exceeding the effects of A alone. They actually refer to situations from which a decrease in the ED_{50} and/or an increase in E_{max} of DRCs can be inferred. It appears that this is the solution of the problem but we do not have to stick to net effects. Observed effects in combination which are greater than the net effects of A in the presence of B represent greater total effects than expected for independently acting agents in combination. The latter, in dose-response studies, are therefore also characterized by a decrease in the ED_{50} and/or an increase in E_{max}, which does not occur in independent combinations (Chap. 2.3).

This approach appears straightforward to define potentiation in complex interaction analogous to simple interactions, namely by a decrease in the ED_{50} and/or an increase in E_{max} – provided that the effect in combination cannot be explained by a sham combination. In other words, a decrease in the ED_{50} and/or an increase in E_{max} can be taken to indicate potentiation in *complex interaction* under the presumption that the agents in combination do not act at the same site. Such a presumption appears justified in most of the situations in which the phenomenon of additivity is associated with a decrease in the ED_{50} in DRCs, as will be discussed in Chap. 3.3.2.

Total effects
Potentiation defined in this way then represents greater effects in combination than expected for independent actions, as already pointed out. Potentiation or synergism between A and B, defined in this way, can thus be characterized by an enhancement of the effects of A by B, i.e., by reduction of ED_{50} and/or increase of E_{max} of A by B. Hence, this definition in complex interaction is in line with the definition of potentiation/synergism and antagonism in simple interaction (Chap. 1.4.1). There are discrepancies of this "effect approach" with the "dose approach", described below. So, it

appears a doubtful procedure to consider additivity as the reference for differentiating synergism and antagonism, as invented by Loewe and Muischnek (1926) and followed by many investigators.

3.1.2 The "dose approach" (evaluation of isobolograms)

The present dilemma exists because quite many investigators follow the approach by studying the *doses* for a given effect *and* consider dose-additive combinations as those which neither show synergism nor antagonism. Then, logically, overadditive combinations are regarded to represent synergism, and underadditive interactions are considered to indicate antagonism, e.g., in isobolograms. Figure 14 illustrates the isobologram approach, and it can be seen that even different investigators somewhat differently interpret isobolograms.

Terms

Some use the terms "overadditive" and "underadditive" as descriptive terms but the majority follows Loewe and Muischnek (1926) and use the terms "synergism" and "antagonism" for equieffective dose-combinations below and above the "additive" dose-combinations (Fig. 14a). Within the

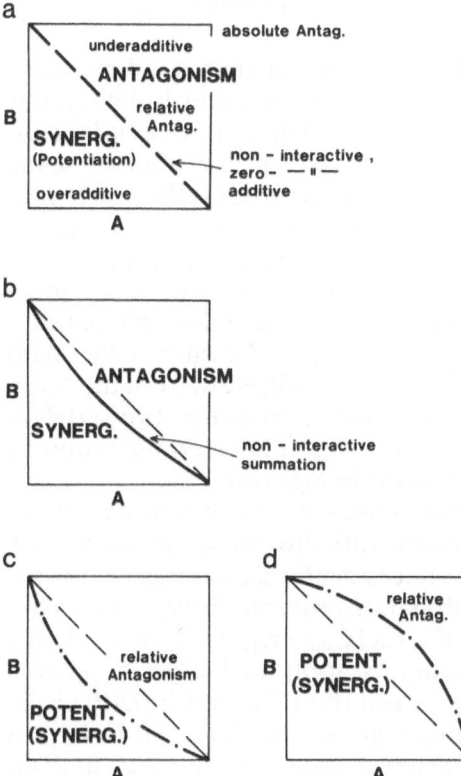

Fig. 14. Comparison of different characterization of synergism and antagonism by schematic isobolograms. **a** The additivity line for "zero-interaction" separates synergism from antagonism, **b** "non-interaction" is considered the reference for mutually nonexclusive drugs, **c, d** calculated independent isoboles (broken dotted line) separate potentiation (*POTENT.*) from relative antagonism, illustrated for a dose-effect relationship with a slope of 0.85 (**c**) and 2 (**d**)

group which uses descriptive terms, "supra-additive" or "superadditive" is preferred over "overadditive", and "subadditive" is used instead of "underadditive" by some authors. Within the group which uses the mechanistic expression, differences may be also noted. Some use the term "potentiation" instead of synergism (Fig. 14a), and some of them use the term "synergism" in a much broader sense, including the range of dose combinations which others characterize as "relative antagonism" (Loewe 1953, Calamari and Alabaster 1980). The latter indicates underadditive combinations in which the doses have to be increased above those for additive combinations, yet not above the equieffective doses of the components acting singly.

If we consider overadditive combinations as synergism and underadditive interactions as antagonism phenomenologically, as many do (e.g., Berenbaum 1989), we have to assume that additive combinations represent neither synergism nor antagonism. Following this line, additivity is considered to represent *zero-interaction* or *no interaction* (Berenbaum 1989, Sühnel 1990, Bödeker et al. 1990). Thereby, a mechanistic criterion is brought into play; phenomenologically we would have to speak of combinations coinciding or resembling additivity. Present day knowledge implies that additivity cannot be regarded as no interaction. Even the same molecules in a sham combination compete for the same binding site(s). Other authors take *no interaction* for those combinations falling exactly on the upper and right boundaries of the square in the isobologram (Sprague 1970, Dawson 1991).

Chou and Talalay (1984) have introduced the concept of independent action into the isobolographic evaluation. They interpret isobolograms either according to the scheme in Fig. 14a *or* Fig. 14b, depending on whether they consider interactions as *mutually exclusive* or *mutually nonexclusive*. The concept of exclusive and nonexclusive drugs is the concept of additivity and nonadditivity, the latter in the sense of independent action. However, the so-called "summation-isobole", claimed by Chou and Talalay (1984) to indicate the isobole for nonexclusive drugs only reflects independent isoboles (see Pöch et al. 1990c) in cases where the drugs exhibit DRCs with a slope of 1 (Christensen and Chen 1985, Gessner 1988). With DRCs of slope < 1 or > 1, independent isoboles are different from the "noninteractive isobole" in Fig. 14b. This fact is evident from Fig. 14c and d, in which the independent isobole reflects an overadditive combination in Fig. 14c, and an underadditive combination in Fig. 14d.

Irrespective of the mentioned differences in the interpretation of isoboles, there are other basic problems with the "dose approach". Although we can assume that lower doses necessary for a given effect indicate *effects* exceeding those of additive combinations (overadditive), we have no information as to whether or not the ED_{50} (at $[E_{min} + E_{max}] / 2$) of A in DRCs is reduced by B! The same considerations apply to higher doses and underadditive combinations, whether or not the ED_{50} of A in DRCs is increased by B! Isoboles do not reflect changes in the ED_{50} and in E_{max} of *DRCs*, despite the fact that changes in *doses* can nicely be seen, e.g., in ED_{50}-

isobolograms. However, the ED_{50} in isobolograms refers to the same effect *level*, whereas in DRCs, the ED_{50} is the half-maximum effective dose (at $[E_{min} + E_{max}] / 2$), which may correspond to an effect level different from 50% (see also Chap. 1.2.1). For instance, the ED_{50} of A may be 1 in the absence of a fixed dose of B, and 0.2 in the presence of B, which may raise the baseline to 60% of the maximum. Then, the altered ED_{50} is obtained at 80%, if E_{max} is 100%.

Let us examine experimental examples. Figure 15 clearly shows that the deviation from additivity does not reflect the magnitude of enhancement of effects. This figure shows the lethal interactions between a benzodiazepine (flurazepam) and ethanol. DRCs of flurazepam alone and in the presence of a fixed dose of ethanol are shown in Fig. 15a; DRCs of ethanol alone and in the presence of flurazepam are given in Fig. 15b. These

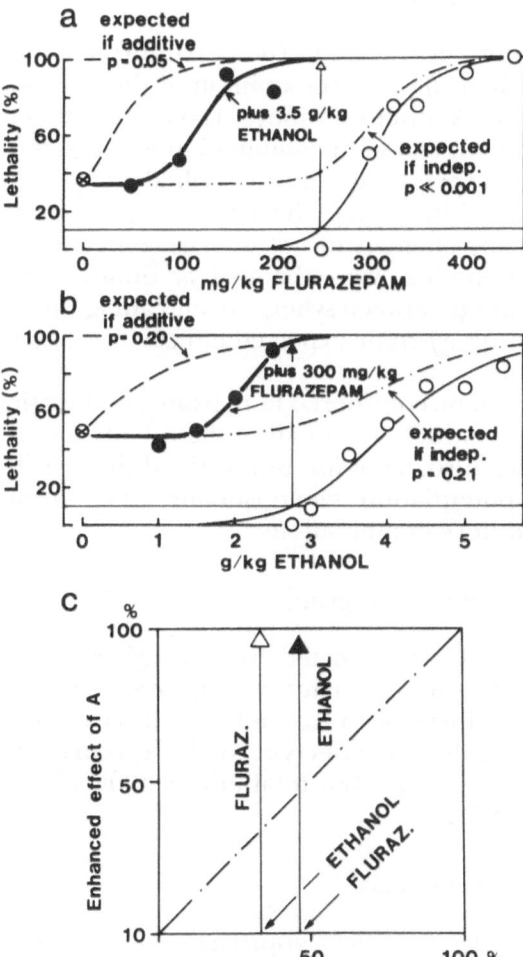

Fig. 15. DRCs (a, b) and combined-effect plot (c). DRCs: enhanced lethal effects in mice of flurazepam (a) and of ethanol (b) in combination. Slope values of DRCs of flurazepam and ethanol, applied singly, were 11.9 and 3.9, respectively. Redrawn from Pöch et al. (1990b). Combined-effect plot: enhancement of effects at the 10% effect level (c), as illustrated by the arrows in the DRCs

experimental results, together with those of other flurazepam-ethanol combinations were first published by Hu et al. (1986) in the form of an isobologram, indicating a clear underadditive combination.

Despite the fact that the combined effects in Fig. 15 appear underadditive in the DRC analysis, they represent markedly enhanced effects in combination. Whereas flurazepam alone at a dose of 250 mg/kg caused only 10% when applied alone, lethality was increased to 100% in the presence of ethanol (Fig. 15a). Also, lethal effects of ethanol were increased from 10% to 98% (Fig. 15b). These increases (of A) are shown in Fig. 15c as a function of the effect of the potentiating agent (B) by arrows, more clearly showing the enhancement of lethal effects of flurazepam by ethanol and vice versa. Hence, underadditive combinations do not in general implicate that the effects in combination are diminished!

3.2 Towards a uniform characterization of synergism/potentiation and antagonism

Consideration of the basic definition of synergism (potentiation) and antagonism as enhanced or diminished *effects*, in the opinion of the author, points to the way out of the dilemma. As discussed above, a combined effect which corresponds to an independent action in combination is characterized by unaltered net effects of A in the presence of B, as described in Chap. 2.

Hence, any deviation of total combined effects from those expected for independent combinations and any changes in net effects, respectively, indicate that the effects in combination are enhanced or diminished. Therefore, we can base differentiation between synergism and antagonism on the "effect approach", i.e., on *net effects* or expected effects of *independently acting agents* in combination.

The sum of individual effects will be considered here from a pragmatic point of view for those special cases which do not allow the calculation of net effects or independent effects. Also, a comparison with additivity will enable us to see in which cases potentiation/synergism and antagonism correspond to over- and underadditive combinations.

3.3 Potentiation (synergism)

Increased *net effects* in combination are reflected by *total effects* which exceed *independent* interactions. They are associated with a decrease in the ED_{50} (at $[E_{min} + E_{max}]/2$) and/or an increase in E_{max} in DRCs. It is a matter of personal preference to evaluate either net effects or total effects in combination. Net effects are easier to interpret but total effects allow also a comparison with additive combinations.

3.3.1 Net effects

Let us first describe and discuss the net-effect approach. As described above, unchanged net effects characterize combined effects of chemical

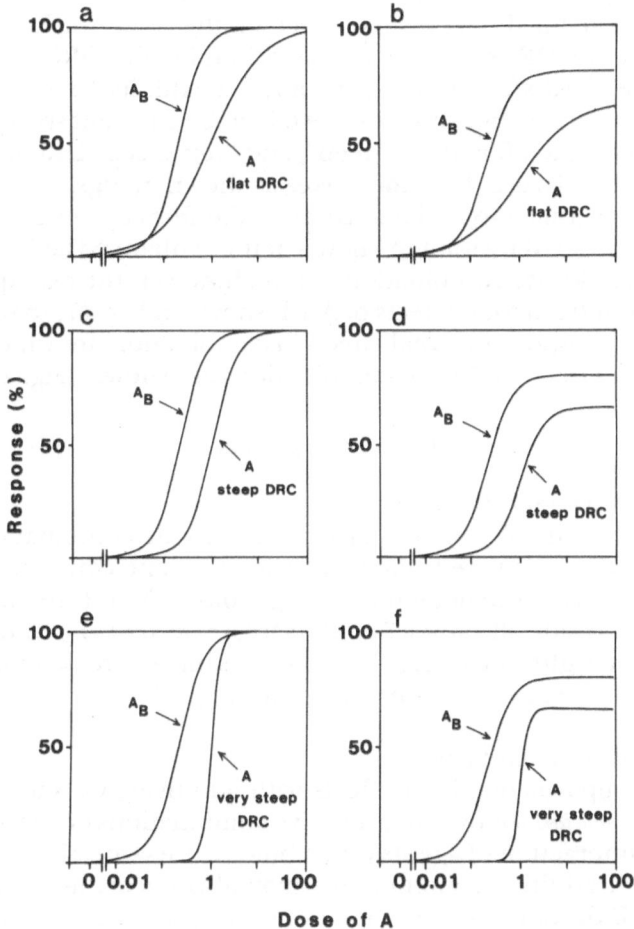

Fig. 16. Enhanced effects of A in the presence of B, expressed by net effects of A_B. DRCs of A_B are all shifted to the left by a dose ratio of 5, with (**a, c, e**) or without (**b, d, f**) an increase in E_{max}. The DRCs of A alone exhibiting different steepness correspond to DRCs of A alone in Figs. 7 and 11. The DRCs of A_B represent net effects which were derived from total effects of A + B in Fig. 17

substances or physical factors when they act independently of each other. Let us first look at idealized DRCs of enhanced actions in Fig. 16. DRCs of enhanced action were assumed to represent a 5-fold decrease of the ED_{50} and/or an increase in E_{max} from 66 to about 80% (Fig. 16), which on the basis of total effects represents an increase from 66% to 88% in Fig. 17. Although Fig. 16 shows idealized curves, we can expect similar curves of combined effects with real data, as indicated by practical examples shown later.

The DRCs of A_B in Fig. 16 thus represent enhanced net effects, with a shift to the left from the DRCs of A alone by a dose factor of 5. The DRCs of A in Fig. 16a–f differ in two respects. In Fig. 16a, c, and e, E_{max} of A is identical with the maximal possible effect, on the right side $E_{max} = 66\%$. In Fig. 16a and b the DRCs of A are characterized by "flat DRCs", in Fig. 16c and d by "steep DRCs", and in Fig. 16e and f A exhibits "very steep DRCs". These six examples in Fig. 16a–f can be taken to represent six typical situations, two curves each for "flat", "steep", and "very steep" DRCs. The DRCs of A alone in Fig. 16 are the same curves as shown in Figs. 7 and 11.

It is a disadvantage to evaluate combined effects by net effects of A in combination with B if we want to know whether enhanced actions of A also represent overadditive combinations. Here, however, the examples of Fig. 16 were derived from total effects of A + B shown in Fig. 17, so we can take a look at the corresponding total effects in combination, for which DRCs of independent as well as additive combinations are shown (Fig. 17).

3.3.2 Total effects

Comparison with independence

In agreement with the theory, all total effects in combination appear greater than the calculated effects of independent combinations (Fig. 17a–f), analogous to the net effects in Fig. 16a–f. Therefore, the DRCs of A + B on the left side all show a 5-fold decrease in the ED_{50}, and the DRCs of A + B on the right side of Fig. 17 show the same decrease in the ED_{50} as well as an increase in E_{max}, comparable with Fig. 16.

Comparison with additivity

Before we compare combined effects with additivity, we should keep in mind that the *mechanism* by which additive combinations occur is caused by competitive interaction of agents which bind to the same, single molecular site. However, additive combinations may also occur, as a *phenomenon*, when A and B do not interact at the same site (see Chap. 2.1.4). "Coincidental" additivity is expected with agents exhibiting a greater steepness than those which compete for a single site. Competitive interactions are typically associated with drugs which exhibit slopes of DRCs around or slightly above 1 (e.g., Pöch et al. 1992).

However, there are situations in which DRCs show slope values greater than 1, described in Chap. 1.2.1, i.e., with molecular or functional cooperativity. Usually, binding to subunits of cooperating binding sites results in slopes of about 2. So we can reasonably speculate that agents with greater steepness than about 2 are likely to exhibit "functional cooperativity" of various actions, resulting in steeper DRCs, comparable to positive molecular cooperativity. As a matter of fact, steeper DRCs are often observed with mixtures of differently acting drugs than with single compounds (see Chou and Talalay 1981, 1984). For example, the slope of the DRCs of two "mutually nonexclusive" enzyme-inhibitors, o-phenanthroline and ADP, is greater when combined (slope = 1.72) than alone (slope = 1.30, and 1.19) (Chou and Talalay 1981). This interaction is shown and described in more detail in Chap. 8.1. Another argument for steeper DRCs than 1–2 associated

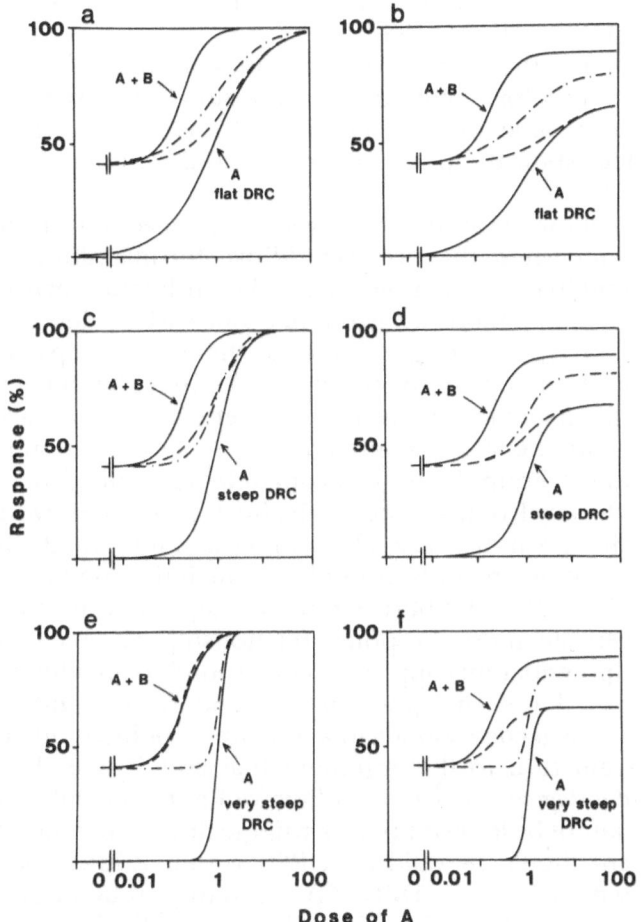

Fig. 17. Enhanced effects of A in the presence of B, analogous to Fig. 16, expressed as total effects of A + B. Expected DRCs for dose-additive (broken line) and for independent interactions (broken dotted line) are also shown to enable the comparison of observed and expected effects. All DRCs of A + B are shifted to the left by a dose ratio of 5 (as in Fig. 16), and exhibit an increase in E_{max} from 80 to 88%, where possible (**b, d, f**) with respect to independent DRCs

with functional cooperativity is the observation that "unspecific" agents exhibit steepness of DRCs with slopes between 1.9 and 3.6 (see Chap. 2.1.4). They nonselectively bind to several sites at the same time.

Only two examples of rather steep DRCs were found in the literature where the compounds studied *in combination* could have interacted at a single, specific binding site. From these examples, slope values of 2.2–2.4 with the herbicides, atrazine and metribuzin (Altenburger et al. 1990), and 2.3–2.8 with the cholinesterase-inhibitors, malathion and parathion (Tammes 1964), could be derived. The herbicides mentioned bind to the D-1 protein of the photosystem II of plants (Trebst 1987), the cholinesterase-

inhibitors to the esteratic substrate binding site of the enzyme, acetyl-
choline-esterase (see Taylor 1990).

It appears that the above given examples are exceptions, since in all
other cases looked for, competitive interactions are rarely expected for
agents with slope values of or greater than 2, often > 1.6 of DRCs, although
they sometimes show the phenomenon of additive interaction (e.g., Hu
et al. 1986).

Figure 17 shows that the observed, enhanced effects are not all
overadditive, and where they are overadditive, the magnitude of the deviations is different in these examples. The relationship between additive and
independent combinations is, as pointed out earlier, dependent on the
steepness and on E_{max} of A (Fig. 12). Hence, the relationship between DRCs
of enhanced action and additivity differs also. In all but one example the
observed enhancement is characterized by effects which clearly outrange
the effects of both theoretical combinations, hence correspond to *overadditive* combinations. From this we can conclude that synergism based on independent effects will be associated with effects of overadditive interactions
in which additivity can be caused by competition for a single binding site.

Figure 17e shows the only example in which the combined effects appear *additive*. The DRC of A here exhibits a steepness with a slope of 5, and
due to this very steep dose-response relationship, greater than independent effects (potentiation) appear to correspond to an additive response,
only. It is likely that such a potentiated response by coincidence rather
than by mechanism corresponds to additivity, as pointed out above.

Let us again turn to the argument that additivity is the *reference for
synergism and antagonism* because additivity of a sham combination of one
compound cannot be looked upon as synergistic or antagonistic with itself.
This statement should be valid for DRCs as well as for isobolograms. In
Fig. 17e we see a shift of the DRC of A + B to the left by a dose factor of 5,
as in the other examples of Fig. 17, yet the DRC of A + B corresponds to an
additive combination. It is theoretically possible that drugs bind to several
identical subunits of an oligomeric protein, hence exhibit (very) steep DRCs,
as in Fig. 17e. This indicates that molecules of one substance can act
synergistically with itself. Hence, the potentiated response shown in Fig. 17e
may occur with compounds binding to *identical* subunits. However, it is
more likely that such a potentiation occurs when A and B bind to different
receptors, simply because a competitive interaction with agents possessing
a slope value of (about) 5 has not been described to the author's knowledge.

The phenomenon of an additive combination associated with very
steep DRCs can then be explained as greater than independent effects,
usually seen with agents acting at different sites (see Chap. 8).

Special situations

In order to express combined effects by net effects it is necessary to know
the possible maximum. Where such a maximum is not known, it is still
possible to evaluate combined effects with respect to the ED_{50} and/or with
respect to E_{max} of a drug effect. Figure 18 schematically shows that a de-

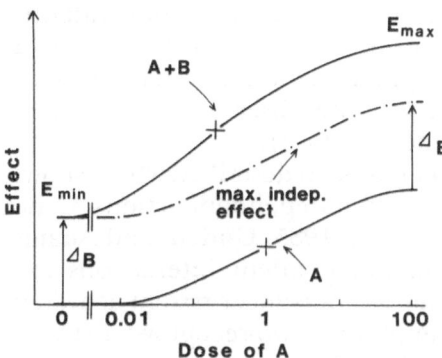

Fig. 18. Schematic DRCs of A alone and in the presence of B. Enhanced effects of A in the presence of a fixed dose of B, which exceed ΔB. The latter represents the maximum possible increase by independently interacting agents (broken dotted line). Enhancement is also evident from the left shift of the ED_{50} (+). The exact DRC for an independent action of A and B cannot be derived because of lack of knowledge of the possible maximum but the ED_{50} of an independent combination is the ED_{50} of A alone (+)

crease in the ED_{50} can be noticed, although the DRC of A + B does not show net effects. This is true for all situations in which complete DRCs of A in the absence and in the presence of B are obtained, as exemplified in Fig. 18.

It is somewhat more difficult to establish an increase in E_{max} above E_{max} of independently interacting agents. Figure 18 again shows schematically how this problem can be solved by calculation of DRCs of the *maximal independent effects*, which represent the *sum of the individual effects* of A and B, also at the maximum of A. This procedure is an application of the effect-addition approach.

The relationship between effect-additive and independent combinations has been shown in Chap. 2.4.2. Briefly, any effect which exceeds the sum of individual effects by necessity has to be greater than the corresponding independent effect. On the other hand, if the combined effects equal Δ_B or are less pronounced, we cannot take such a result to indicate that the observed effects are equal to or less than independent! Hence, we cannot take such a finding as lack of potentiation.

3.4 Antagonism

Again, there is a discrepancy among several investigators about antagonism in complex interactions. Basically, we have the same problems with "antagonism" as with "potentiation" described above – but we can look at antagonism as a mirrorlike phenomenon to potentiation.

3.4.1 The "effect approach"

If we accept antagonism as the antonym of synergism/potentiation and if we characterize synergism by a decrease in the ED_{50} and/or an increase in E_{max}, then antagonism should be defined by the opposite changes. Here, we will not deal with situations in which B has either no effect alone or an opposite effect to A.

From a practical point of view, we can rely on either net effects or on total effects which correspond to independent combinations. So, reduced net effects or smaller total effects than those expected for independently interacting compounds indicate antagonism. A look into the results of complex interactions reveals that the phenomenon of pharmaco- or toxicodynamic antagonism is much less frequently observed than potentiation (with agents acting singly in the same direction).

However, a certain type of interaction has been described which shows a diminished combined effect with respect to independent actions of the components (Black and Leff 1983, Black et al. 1985, Undem and Adams 1988). In practice, the deviations from independent interactions are rather small and are explained by a common transducer system activated via different receptors by A and B. Examples for two-receptor-one-transducer mechanisms (Undem and Adams 1988) will be given in Chap. 8.

Another example of antagonism, resulting in diminished net effects, is due to interaction of a full with a partial receptor agonist, a special type of competitive interaction, described below.

3.4.2 The "dose approach" (evaluation of isobolograms)

In contrast to antagonism defined by the "effect approach", *underadditive combinations*, often understood as antagonism, are much more often seen. Phenomenologically, underadditive combinations may be caused by independent or even greater than independent effects of compounds, acting at different molecular sites. This occurs with compounds exhibiting very steep DRCs (Figs. 13e and 15a–c).

Hence, the phenomenon of underadditive combinations in some, if not many, situations may reflect independent interactions which, on a mechanistic basis, can hardly be ascribed to antagonism. It is the situation depicted schematically in Fig. 14d where the isobole of independent effects corresponds to relative antagonism. Quite a number of examples of isoboles for independent or greater than independent effects could be mentioned here, some of which have been published recently (Pöch et al. 1990c).

Underadditive combinations can occur with drugs acting at the same molecular site, but possessing different intrinsic activities, i.e., different E_{max} values. This type of antagonism between a full and partial agonist is considered to represent a dual mechanism of B, a synergistic and an antagonistic component of action (Ariëns 1956a; see also Pöch and Zimmermann 1988). Here, the underadditive combined effect clearly represents an antagonistic mechanism of B, which is reflected by combined effects which are less pronounced than those expected for independent combinations.

3.5 Comparison of complex with simple interactions

A uniform characterization and definition of the term potentiation (synergism) and antagonism appears desirable and possible on the basis of net effects in combination and total effects of independently acting agents, respectively, described and discussed above. If we accept the characterization of potentiation and antagonism as a change in the ED_{50} and/or in E_{max} of DRCs, it appears reasonable that we can uniformly define these terms in complex and simple interactions.

Even if we do not accept this definition, we could argue that in situations where we can exclude the same binding site, and different sites of action are likely in most of the studies of combined effects, we can dismiss an evaluation with respect to additivity. Then, if we solely analyze the effects with respect to net effects or total effects of independently interacting agents, we actually compare complex interactions in analogy to simple interactions. In the latter, we evaluate effects in combination with respect to the ED_{50} and E_{max} (Chap. 1.4.1).

3.6 Quantitative expression of combined effects

The quantitative expression of combined effects is another point where no agreement is in sight. A simple, descriptive graphical expression has already been mentioned in Chap. 1 (Fig. 3). It will be described in detail with examples in Chap. 7 as "combined-effect graph". It simply shows the magnitude of enhancement or diminution of effects.

Other, more sophisticated procedures express the magnitude of effects by the shift of the ED_{50} of DRCs or by the factor by which the doses can be reduced to obtain the same magnitude of effect in combination, or the like.

3.6.1 The combination index (CI)

Probably the currently most widely used expression is the "combination index" (Chou and Talalay 1984, Berenbaum 1989). It is a quantitative expression of the magnitude of deviation of observed combinations from expected *additive combinations*. CI equals the sum of the fractions of equieffective doses of the components of a combination (Berenbaum 1989) for mutually exclusive agents (Chou and Talalay 1984). It is analogous to the FIC index, the sum of the values of the fractions of equieffective inhibitory concentrations, introduced by Elion et al. (1954).

These indices are certainly useful to express the results of combined effects by a single value with respect to the *doses* of the agents, which are necessary to produce a specified magnitude or level of effect. However, CI values do not reflect the magnitude of enhanced or diminished *effects*. In other words, this index does not (uniformly) indicate the magnitude of enhancement or diminution, as does the combined-effect graph, or as can be derived from a simple inspection of DRCs. Therefore, a quantitative expression of the deviation from *independent effects* appears much more meaningful with respect to combined effects.

Let us consider an example, in which A and B separately produce 25% effect. We state that the combined effect of A plus B would exceed the calculated independent effect of 44%, this means > 44% in this example, irrespective of the steepness of the DRC of A (Table 4).

Table 4. Comparison of independent and dose-additive effects

DRC slope	> Independent (%)	Additive (%)
0.8	> 44	37
1.0	> 44	40
3.0	> 44	73
5.0	> 44	91

As is evident from the comparison with calculated effects of dose-additive combinations, greater than independent effects are greater than additive in situations in which DRCs of A have a slope of 0.8 to 1. They therefore correspond to CI values < 1, discussed above. However, with steeper DRCs, greater than independent effects may be less than those expected for additive combinations, unless they are greater than 73% for a slope of 3, and greater than 91% for a slope of 5, in this example. Hence, CI values > 1 simply reflect underadditive combinations!

3.6.2 Comparison of CI and combined-effect graph

Inspection of the DRCs in Fig. 15 finds enhanced effects exceeding independent effects from the 10% effect level. Hence, it is evident that the combined effects of flurazepam were dramatically enhanced by ethanol and vice versa. Further, the combined effects were greater than expected for independently interacting agents, although they were not greater than those calculated for additive combinations, thus correspond to CI values for underadditive combinations, by definition greater than 1.

Finally, CI values do not tell whether the combined effects are greater or smaller than those of independent interactions since they are based on a comparison with additivity – except were additivity matches with independence (see Chap. 2).

3.7 Suggestions for quantitation of combined effects

As already described and discussed, the magnitude of deviations of combined effects from independent effects can be expressed by the changes of effects. In DRCs we can quantitate deviations by the shift of DRCs, as indicated in Figs. 16 (net effects) and 17 (total effects). In all examples, the assumed DRCs of observed combined effects were shifted to the left by a dose factor of 5, the dose ratio.

In these examples, when combined effects in DRCs are expressed by the total effects in combination, the shift of the DRC of the combination equals 5 with respect to the DRC of independent interaction as well as with respect to the DRC of A in the absence of B. This fact is explained by the circumstance that the ED_{50} of DRCs of independent actions equals the ED_{50} of A alone.

The *practical aspect* of this fact is that for quantitating as well as for diagnosing combined effects expressed by DRCs (total effects), it is not necessary for estimation of the ED_{50} to calculate the respective DRC of independent interaction.

When results are to be evaluated with respect to E_{max}, an increase in the combined effects is immediately seen in enhanced net effects at the maximum of the DRC (see Fig. 16b, d, and f). However, when expressed by total effects and when an increase in E_{max} is possible (Fig. 17b, d, and f), the enhanced maximum of total effects should be compared to the maximum of independent effects, since the latter represents an unaffected response.

3.8 Tips and hints

When we want to analyze DRCs of combined effects with respect to potentiation or antagonism, it is not necessary to express the combined effects by net effects or to calculate independent effects, if we only look for changes in the ED_{50} of A in the presence of a fixed dose of B. We can detect such changes, whenever we compare complete DRCs. Note that the ED_{50} (at $[E_{min} + E_{max}]/2$) in DRCs is the half-maximum effective dose of the respective curve, regardless of the basal level, as indicated graphically in Fig. 18. This example also shows that it is not necessary to know the possible maximum of response of the system in order to determine the half-maximum effective dose, i.e., the ED_{50}.

Any left shift of a DRC of A in the presence of a *fixed dose of B* indicates greater effects than of A alone, i.e., greater absolute effects in simple combinations (see Fig. 2c), greater relative effects in complex combinations (see Figs. 15–18). The latter represent greater than independent effects. Note that about equieffective *mixtures* show a left shift when the components act independently together (see Chap. 4.2).

When reports in literature are studied (and compared to own investigations), we should first check which meaning the authors use for the terms "synergism", "potentiation", "antagonism", etc. Then, we should check whether the conclusions drawn are valid. In this respect, we should keep in mind that many interpretations based on isobolograms are inadequate or even wrong, if overadditive and underadditive is interpreted to mean enhanced and diminished effects in combination, respectively. If in doubt, it is probably better to describe the combined effects without labeling it as potentiation (synergism) or antagonism.

4 Evaluation by dose-response curves

Various models of agonism of drug action have been developed, expressed and interpreted by DRCs (e.g., Ariëns et al. 1956a, b; Black and Leff 1983; Ehlert 1986, 1988). Also, models of synergism and antagonism have been expressed by DRCs (e.g., Van den Brink 1977; Unkelbach and Wolf 1985a, b; Mackay 1981; Plummer and Short 1990). Most of the interaction models were used to study the effects of one agent in the presence of fixed doses of another agent (e.g., Ariëns et al. 1956a, b; Van den Brink 1977).

Under the presupposition that A and B alone phenomenologically lead to a similar response, there is the possibility that A and B behave like dilutions of one another, which in combination agree with the model of additivity. The underlying mechanism, i.e., an action at the same molecular site, has been described in Chap. 2. Another possibility is that compounds with similar effects act at different sites. Their combined effects may or may not agree with the model of independence (Chap. 2.3), as evident from experimental results described in Chap. 8.

We can test for the site of action by comparing the DRCs of observed combined effects of A in the presence of B with DRCs of effects expected in case they share the same site of action, i.e., with additivity. Competition for a binding site results in a dose-additive response or in a weaker response, depending on the ability of the drugs to activate a reaction triggered by binding to the "receptor site". The latter ability is phenomenologically expressed by the magnitude of response, i.e., by E_{max}. In pharmacology this property of drugs is termed "intrinsic activity" (Ariëns et al. 1956a). When B has the same or a higher intrinsic activity as A, the combined effects of A and B via activation of the same receptor-effector mechanism correspond to dose-additive combinations as described and illustrated above (Fig. 6b, d). When one agent has less intrinsic activity than the other, the combined effects of A and B will be less than dose-additive (Fig. 6f), usually termed underadditive or subadditive. An example will be described in Chap. 4.1.3.

The usual approach to study this kind of dual interaction is to do a dose-response experiment of the "full agonist" alone and in the presence of the "partial agonist" (Fig. 6f). If the full and the partial agonist compete for the same receptor, an underadditive response is expected, i.e., a parallel shift of the DRC of A + B observed to the right from the theoretical additive DRC (see Pöch and Zimmermann 1988), with no change in E_{max} of the full agonist. If a full agonist is tested in the presence of another full agonist of the same receptor, an additive response is obtained (Fig. 6b), and this response shows no change in E_{max} either.

However, if E_{max} of A + B exceeds E_{max} of A, as indicated in Fig. 19c, then we can assume that the two agents studied in combination act at different sites. Hence, different sites of action of A and B are likely if E_{max} of A is significantly increased by B and if the observed response of A + B significantly exceeds the expected dose-additive response, respectively. In this situation, we might be interested either in the magnitude of the combined effects and/or in the mechanisms by which this interaction occurs. As has been pointed out in the foregoing chapters, the outcome of experiments with agents acting at different sites, leading to a common response, may or may not correspond to independent actions of the components.

Only *independent actions* of agents acting at different sites can easily be calculated, as described, but other interactions are also logically expected for agents acting at different sites. Thus, a comparison of observed with expected independent interactions may suggest how the agents under study actually interact, whether potentiation (synergism) or antagonism occurs, described in Chap. 3.

Furthermore, a comparison with independent interaction is also of some interest for site-directed analysis, since combinations which are not

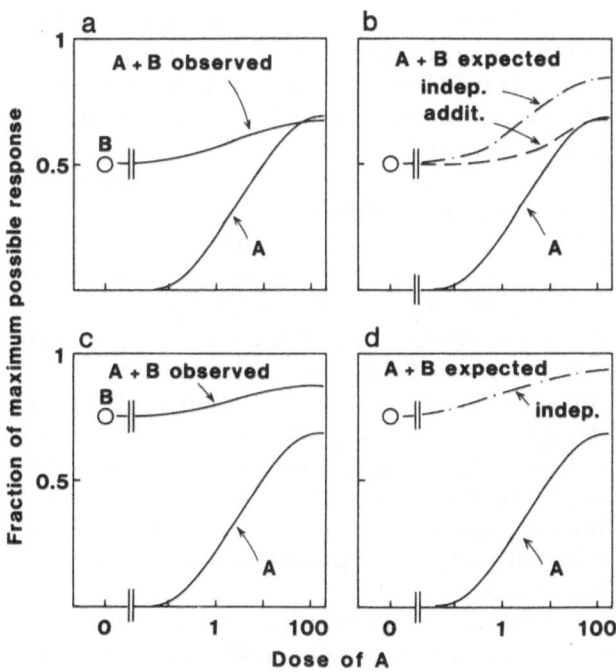

Fig. 19. Methodological aspects. Schematic DRCs of A alone and in the presence of B. **a, b** The "observed" effect of B is lower than E_{max} of A, hence additive (broken line) as well as independent (broken dotted line) DRCs can be calculated (**b**). **c, d** The effect of B exceeds E_{max} of A, hence only the respective independent DRC can be calculated (**d**)

significantly different from independent interactions logically do not exclude different sites of action. So, if an interaction appears not significantly different from additive or from independent interaction, no safe conclusions can be derived from such an experimental result. This situation is not rare. Basically, such results may be explained in two ways. Agents which act at different sites often show combined effects similar to the one expected for independent actions (see Chap. 8). If the DRCs of independent and additive interactions are not well separated, it may be impossible to differentiate between independent and dose-additive combinations. The former suggest different sites, the latter "allow" the same site of action. Where the two theoretical DRCs appear nicely separated there may still be the situation that combined effects are not significantly different from either one, because of insufficient number of experiments or large variation in results.

4.1 New methodological approach: fixed-dose studies

From the above considerations a new methodological approach can be derived which implicates the evaluation of combined effects by direct analysis of DRCs of A alone and in the presence of a *fixed dose of B*. For this purpose, the corresponding theoretical DRCs for dose-additive and independent combinations, respectively, have to be constructed in order to allow a comparison of observed with expected responses. The principle of this approach is schematically illustrated in Fig. 19.

This approach is considered new inasmuch as it represents an alternative to procedures evaluating *mixtures* of A and B, either by analysis of DRCs or by isobolograms. Studying mixtures of A and B, i.e., fixed dose-*ratio* combinations offers some advantages, besides a number of drawbacks, described in Chap. 4.2 from a pragmatic attitude.

The method described for the evaluation of the effects of A in the presence of a fixed dose of B has only quite recently been described for frequency studies and other experiments with respect to a solid and practicable statistical analysis (Pöch et al. 1990a, b). We will use the expression "fixed-dose studies" for this procedure. Advantage was taken also of the possibility to fit experimental points to sigmoidal DRCs (Pöch et al. 1990a). The new approach is applicable for all studies in which DRCs can be obtained. On the basis of this procedure, theoretical DRCs for dose-additive or independent actions can be predicted from DRCs of A and effects of B alone, the latter either measured or assumed. As a matter of fact, this possibility already was exploited for the construction of theoretical DRCs in Figs. 6 and 7.

The practical implication of assuming a certain response to B alone is to check whether or not the theoretical DRCs for additivity and independence will be separated from each other at that effect level of B. For example, no separation between dose-additive and independent DRCs is evident in the situation depicted in Fig. 13c. In other situations, additive interactions are reasonably below (Fig. 13a at an effect of B = 0.6) or above the

DRCs of independent actions (Fig. 13e). Hence, there are limitations of this method which can be checked beforehand. However, where we can expect greater effects in combination than with both, additively as well as independently interacting drugs, lack of differences between the theoretical curves may be no handicap. As is implicated in Fig. 13a and b, we may be able to choose optimal conditions for a planned study, i.e., the best possible separation of the theoretical curves when the results are expected to correspond either to additive or to independent interactions.

4.1.1 Construction of experimental and theoretical dose-response curves

Whenever possible, the dose response of A alone and in the presence of B should be studied over the full effective dose range to allow construction of complete DRCs. The construction of the curves themselves can nowadays be done with the aid of a proper curve-fitting program, e.g., ALLFIT (De Lean et al. 1988). Thereby, experimental as well as theoretical DRCs of dose-additive or independent interactions can be constructed. The basis of constructing theoretical DRCs has been given in earlier chapters, and schematically illustrated in Figs. 5 and 10. Methodological details for the construction of theoretical DRCs have been described recently (Pöch et al. 1990a).

Practical approach and limitations
In Appendix B of this book, a simplified approach (unpublished) will be described, and examples for practical examination by the reader are supplied. The practical construction of theoretical DRCs *rests* on the following considerations. The effects of *dose-additive* interactions can simply be derived from the dose-effect relationship of A alone: the same effects as obtained with A alone can be expected at the doses of A minus the dose x of A, with which B is equieffective (Fig. 5). For instance, B is equieffective with the dose of 2A, i.e., x = 2A. Then, for example, the effect at the dose of 2.5 A (62%) is expected in an additive combination at the dose of 2.5 minus x, i.e., at the dose of 0.5A (Fig. 5).

If B alone would have shown an effect of 45%, then, in this example, the *independent* effect at a dose which elicits 62% (2.5A) would amount to 79% (Fig. 10). This value is obtained by converting percent values into fraction-of-maximum values and the independent effect in combination then is calculated by Eq. (1) from

$$0.45 + 0.62 - (0.45 \times 0.62) = 0.79.$$

If the effect of B exceeds the maximum effect of A (Fig. 19c), only DRCs for independent combinations can be constructed as illustrated in Fig. 5. If the maximum possible response is not known, no DRC for an independent combination can be constructed. The latter limitation has been discussed in Chap. 3.3.2.

Quantitative and qualitative studies

In principle, experimental data of quantitative studies are treated different from quantal or qualitative studies, i.e., frequency experiments, in order to allow proper statistical analysis. It is clear that frequency studies can be expressed by the percentage of reactants, quantitative studies by the fraction of the maximum. There is no need to express frequency data by mean or median values. In quantitative studies, we calculate median rather than mean values to allow a statistical analysis by chi-square goodness-of-fit, to be described below.

Multiple combinations

More than two agents can be studied in combination by analogy with pairs of drugs by testing the dose response of A in the presence of a mixture of other agents. Also, the results of such experiments can be analyzed analogously, taking into account that A was tested in the presence of a mixture. As a matter of fact we often test, without intention, compounds in the presence of a mixture, namely a racemic mixture.

4.1.2 Statistical comparison of observed and expected response

Chi-square goodness-of-fit test

In *frequency studies*, we can directly compare the number of reactants observed with the number expected. The latter depends on the total number of observations. For instance, as shown in Fig. 20a, at the dose of 0.6 ppm parathion the number of killed flies was 27 out of 50. At this dose 36% would have been expected dead, which means 18 out of 50, if parathion and malathion had interacted independently of each other. Figure 20a shows this comparison over the observed dose range from 0.6 to 5 ppm parathion. The observed numbers of reactants are then compared with the expected numbers over the entire dose range by the chi-square goodness-of-fit test. In this example, chi square (χ^2) was estimated to equal 17.04 with 3 degrees of freedom (d.f.), from which the value of 0.0006 for p was derived.

In *quantitative studies*, we can dichotomize the values of response by expressing experimental results by median values, from which theoretical DRCs can be derived which also represent median values. Since median values split the number of individual values into half above and half below them, it is possible to check for statistical deviations of observed from expected values by comparing values above or below the theoretical median values. The expected values are always half of the values under consideration. Intuitively, we could think that this should be true for mean values also. However, this is mathematically correct only in case of a normal distribution, which in usual practice cannot be proven. Figure 20b illustrates such a comparison of observed with independent interactions. This experiment was chosen because of the relatively small number of experiments which allows a reasonably clear graphical illustration.

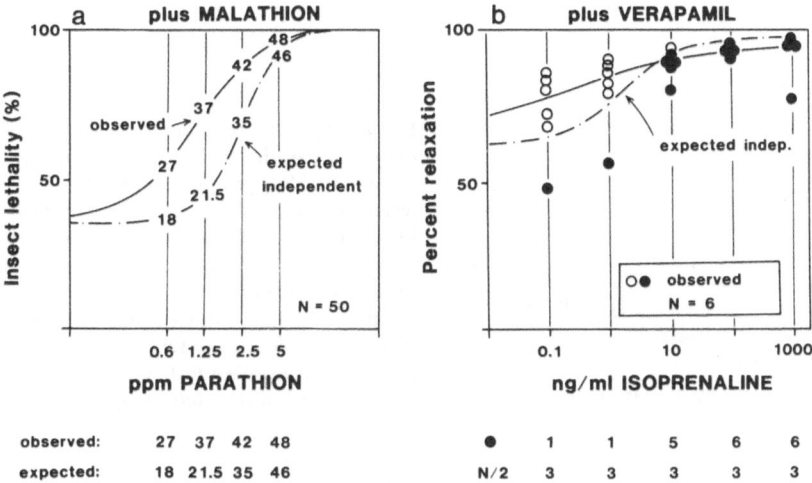

Fig. 20. Illustration of statistical comparison of observed and expected response in combination in frequency studies (**a**) and in other studies (**b**). The examples represent actual experiments of insecticidal effects (**a**) of cholinesterase-inhibitors (see Fig. 21) and of isolated organ experiments (**b**) with bovine tracheal muscles (see Pöch et al. (1990a) for methodological details). Malathion was applied at 20 ppm (**a**), verapamil at 20 ng/ml (**b**). Comparison of "frequencies" by χ^2 goodness-of-fit "observed" vs. "expected". N equals the total number tested at each datum point, N/2 half the number of observations, which are expected above and below the calculated curve of independent action in **b**. ● Values below, ○ values above the calculated median DRC of an independent interaction. Results showed significant differences compared with expected independent effects in both graphs (**a**, p < 0.001; **b**, p = 0.04). See also Appendix B

F-test

De Lean et al. (1987, 1988) invented what they called the F-test to check whether DRCs could be fitted with a certain constraint without significant loss of fit. The principle is that the data points (mean values) are fitted first without, and then under constraint, e.g., of same slope. This procedure can also be used for the statistical analysis of DRCs of combined effects. A first fit is run without, and a second fit by forcing experimental values to theoretical curves. This test appears very "sharp", i.e., detects rather small deviations with p < 0.05. Some examples are given in this book. It has not been used for statistical analysis of DRCs when points were obtained by cumulative dosing of isolated muscle strips.

4.1.3 Expression of results and examples

There are many ways to express results of combined effects, however, here we follow a scheme that has proven useful and adequate. Routinely, we express the effects of A alone by open symbols, combined effects of A plus B in DRC analysis by filled symbols, which represent frequency values or

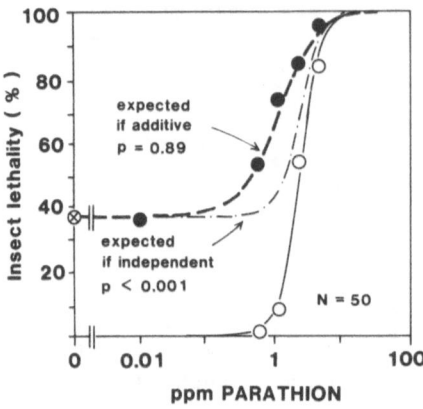

Fig. 21. Quantal response. Insecticidal effects of parathion alone (○) and in the presence of 20 ppm malathion (●). Data were taken from Tammes (1964) and fitted by ALLFIT. Slope of the DRC of parathion alone was 2.8, the effect of parathion in the presence of malathion appeared dose-additive (p = 0.98) and significantly different from independent interaction (p < 0.001) in the χ^2 goodness-of-fit test. The latter comparison is illustrated in Fig. 20

median values of quantitative studies. *Individual* values of the latter DRC studies are only given in Fig. 20 to illustrate the analysis of combined effects by chi-square goodness-of-fit. Also, for clarity, the DRCs for the theoretical DRCs are uniformly indicated by broken lines for dose-additive inter-actions, and by dash-point or dotted lines for independent actions throughout this book. With a few exceptions, the DRCs are presented as upward curves. Examples for examination are provided in Appendix B.

Frequency studies
As an example of frequency studies, the insecticidal effects of cholinesterase-inhibitors are depicted in Fig. 21. It shows the effects of parathion alone and in the presence of a medium effective dose of 20 ppm malathion in comparison with the theoretical DRCs for dose-additive and independent interactions expected. The two theoretical DRCs were reasonably separated at the effect level of 35.5%, achieved with 20 ppm malathion alone. The differences between the *theoretical DRCs* were less at lower doses of malathion (not shown). From the results of this experiment it can be concluded that the two cholinesterase-inhibitors might have killed the flies by a shared action at the cholinesterase (not significantly different from additive) rather than by acting at different sites (significantly different from independent).

Another form of presentation has been adopted to illustrate the effects of frequency with more than one dose of B, to be exemplified in Chap. 8. This approach was also used to compare observed, quantitative effects with theoretical effects of additivity and independence. There, for reasons of graphical clarity, the observed DRCs are presented in one graph and the theoretical DRCs for additivity and independence, respectively, in separate graphs.

Quantitative studies
Our usual way of presenting data is to show combined effects of two drugs together with the theoretical DRCs (Pöch et al. 1990a). Figure 22 shows results of drugs acting at the same site expressed in this way. It can be seen

that in these examples the effects in combination correspond to the median effects of dose-additive combinations. The differences between observed and calculated effects were not statistically significant ($p = 0.25$ and
$p = 0.99$, respectively). On the other hand, the observed effects clearly and
significantly ($p = 0.01$ and $p < 0.03$, respectively) differed from calculated
independent interactions. So, we can be quite sure that these drugs inter-

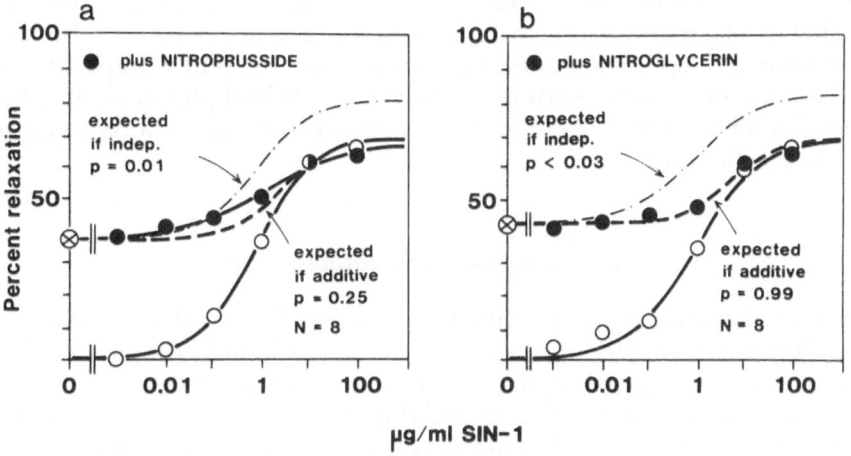

Fig. 22. Quantitative response. Examples of combined effects with drugs interacting
additively in relaxing isolated smooth muscles. Results of experiments with bovine isolated
coronary arteries (**a**) and tracheal muscle (**b**), precontracted by 27 mmol/l KCl. Slope of the
DRCs of SIN-1 (○) were 0.75 (**a**) and 0.65 (**b**), respectively. The combined effects, median
values (●), were not significantly different from additive but were significantly less pronounced than expected for independent interactions ($p < 0.03$) in the χ^2 goodness-of-fit test.
Redrawn from Pöch et al. (1990a)

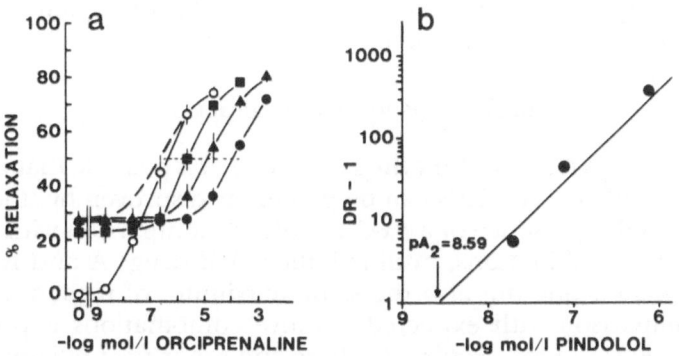

Fig. 23. Quantitative response. Example of drugs interacting underadditively in relaxing
smooth muscles. **a** DRCs of the full agonist, orciprenaline, alone and in the presence of the
partial agonist, pindolol, applied at concentrations of 18 (■), 80.5 (▲), and 805 (●) nmol/l,
with additive DRC for 80.5 and 805 nmol/l pindolol (broken line), **b** Schild-plot of antagonistic action. Redrawn from Pöch and Zimmermann (1988)

act at the same molecular site since the statistical treatment does not exclude the same site but excludes different sites of action.

For comparison, we also show the effects in combination of drugs acting at the same site, the b-adrenergic receptor, but as full (orciprenaline) and partial agonists (pindolol) in Fig. 23. Here, the two agonists do not behave like dilutions of one and the same substance, yet they bind to the same receptor site. The DRCs of orciprenaline are dose-dependently shifted to the right in parallel to the additive DRCs. The antagonistic component of this dualistic interaction is evident from this parallel displacement from the corresponding additive DRCs. The latter represents the agonistic component of interaction. The competitive nature of this interaction is strenghtened by the Schild-plot analysis, which yielded a straight line, expected for competitive interactions following the mass action law.

4.2 Fixed dose-ratio combinations

Mixtures of agents applied in fixed dose *ratios* of A and B have been used quite frequently in the literature for the evaluation of combined effects, mainly for isobolographic analysis. In some studies with mixtures, the DRCs have been analyzed directly and the observed DRCs of a combination of A and B have been statistically compared with additivity and/or independence (Unkelbach and Wolf 1985a). It is important to realize that additivity has to be considered in this approach as pure phenomenon, since it has been calculated regardless of whether or not the individual DRCs showed the same slope or not. From the mechanistic point of view described above and followed in this book, such an approach is not appropriate for substances with DRCs which significantly differ in slope. In the latter case, A and B appear to behave differently. As a matter of fact, most investigators consider different slopes as indicator of differences in the site of action, although this assumption, to the knowledge of the author, has not been investigated thoroughly so far.

4.2.1 Dose-additive interactions

However, if drugs A and B show the same slope, it is possible that they share the same site of action. (This can be proven or disproven by studying the DRCs of A in the presence of a *fixed dose of B* by comparison with effects of dose-additive combinations, outlined above.) If drugs A and B not only share the same slope but also the same maximum of response, a rather simple comparison with expected additive combinations is possible, as outlined in Fig. 24a for *mixtures*. In these and other *fixed-ratio* graphs dose scales for A and B (and C ...) are given separately, rarely seen in the literature (e.g., Hofmann et al. 1989, Liew et al. 1990). These dose scales are superimposed, so to say, and reflect the applied dose ratio. This, in turn, will aid the construction of the theoretical DRCs and the understanding of

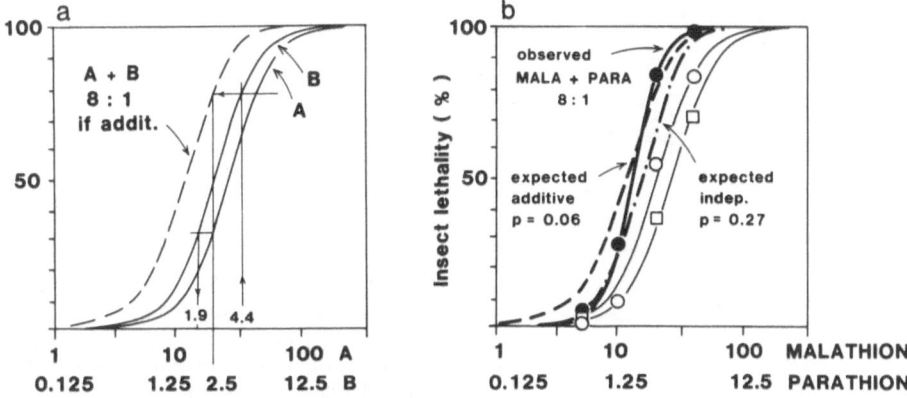

Fig. 24. Schematic and experimental fixed-ratio combinations. **a** Schematic DRCs of A and of B alone and combined in a fixed dose ratio (8:1). The DRCs of A and B exhibit the same slope and the same maximum. Hence, the theoretical DRC for an additive interaction can meaningfully be constructed as a parallel DRC, whose shift to the left was determined at the dose of 2.5 B. A at that dose was equieffective with 1.9 B, hence the dose-additive response was that at the dose of 2.5 + 1.9 = 4.4 B. **b** Experimental fixed-ratio combination of malathion and parathion (●) in comparison with the insecticidal effects of the mixture components alone (malathion □, parathion ○) and in comparison with the theoretical DRCs for dose-additive (broken line) and independent actions (broken dotted line). The DRCs of malathion and parathion exhibited slope values of 2.2 and 2.8, respectively, and were fitted to the same slope (F-test: p = 0.09) by ALLFIT. The response to the mixture was neither significantly different from an additive (p = 0.06) nor from an independent interaction (p = 0.27) in the χ^2 goodness-of-fit test

experimental results of fixed-ratio combinations, illustrated in Fig. 24b. Otherwise, it is quite difficult to predict where the DRCs of combined dose-additive or independent action will be.

The interaction between the cholinesterase-inhibitors, malathion and parathion, has been shown in Fig. 21 by DRCs of the insecticidal effects of parathion alone and in the presence of a fixed dose of malathion. From the same set of data (Tammes 1964), DRCs of the effects of a fixed-ratio combination of malathion and parathion were constructed and compared with additive and independent interactions (Fig. 24b). The effect in combination was neither significantly different from independent action (p = 0.27) nor from additivity (p = 0.06). It is interesting that the DRC of the mixture shows a greater steepness than the DRCs of malathion or parathion alone, and the parallel, additive DRC, respectively, however, this difference was not significant in the F-test (p = 0.06).

Under the presupposition that the interacting agents exhibit the same slope and the same maximum, as in the example of Fig. 24, the constructed additive DRC is the expected DRC of agents interacting at the same site. If the *maxima* are *different*, we have to expect a dualistic interaction between the agonists with higher and lower intrinsic activity, which correspond or imitate the interaction of full and partial receptor agonists. The expected

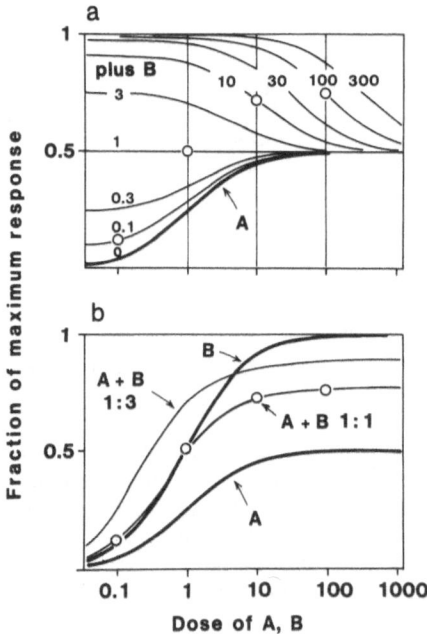

Fig. 25. a Calculated DRCs of a partial agonist (A) alone and in the presence of a full agonist (B) with equal affinity, showing competitive dualism (Van den Brink 1977). ○ The calculated reponse in combination (A + B) at a fixed dose ratio of 1:1. **b** DRCs of A and of B alone and combined as mixture of A + B = 1:1, and 1:3, respectively, are shown. The effects of the mixtures were derived from **a**. The ordinate of the graph is for A alone, B alone, and the mixtures in terms of the dose of A contained in the mixtures

DRC of a partial agonist A in the presence of a full agonist B can be derived from model curves of Van den Brink (1977), depicted in Fig. 25a.

In Fig. 25a the DRCs of A in the presence of 0.1, 0.3, and 1 B correspond to dose-additive curves. They represent the synergistic part of the dualistic action between A and B. In the example of Fig. 25 both drugs are assumed to act at the same site but with different intrinsic activities. Hence, at concentrations of B with effects above E_{max} of the partial agonist A, i.e., above the dose of B = 1, the antagonistic component becomes evident in Fig. 25a. The competitive type of interaction of A and B at the same receptor is seen as parallel shift of the curves in the presence of 10 to 300 B (Fig. 25a).

This parallel shift is not seen in fixed-ratio combinations (Fig. 25b) which can be derived from the graph in Fig. 25a, e.g., for a 1:1 ratio of concentrations, the effects of which are indicated in Fig. 25a. In order to illustrate the problems with mixtures, the effects of a 1:3 mixture are also shown in Fig. 25b, derived from Fig. 25a.

The resultant response cannot as easily be derived as by following the fixed-dose study approach. So, if we are not only interested in phenomena but also in mechanisms by which substances in combination interact, we have to restrict the evaluation of fixed-ratio combinations to experiments where A and B show the *same slope and the same maximum* or we have to use an appropriate equation, yet to be developed, to model this interaction. Alternatively, we can avoid running into trouble with fixed-ratio combinations by replacing studies of mixtures by the fixed-dose approach described above (Chap. 4.1). Examples of this are also given in Chap. 8.

4.2.2 Independent actions

It has been pointed out above that the evaluation of mixtures offers some practical features. Whenever the *magnitudes* of combined effects of mixtures are of interest, e.g., in clinical pharmacology or in ecotoxicological studies, fixed-dose ratio combinations appear appropriate. The combined effects of the mixtures can be compared with expected effects of independently interacting components of mixtures, regardless of their slope and E_{max}. This means that we could always analyze the effects of mixtures with independent combined effects – as long as we know the possible maximum of response. If not, we can, by substitution, use the effect-additive approach described and illustrated above (Fig. 18).

The calculation of independent combined effects of mixtures can be carried out rather simply (Fig. 26a). We only have to take the individual values of effects of A and of B at selected doses for calculation of independent combined effects, described in detail earlier (Chap. 2). At the doses of 10A and 1B, calculation of independent effects is exemplified in Fig. 26a. The point for the DRC of independent action is derived by calculating

$$0.6 + 0.8 - (0.6 \times 0.8) = 0.91.$$

There is an important point that shall be stressed here. In a response-oriented study in which we are interested in the magnitude of combined effects of mixtures, it appears of great advantage to present the DRCs of A and B alone, and in combination by using *separate dose scales* for the agents combined at a fixed-dose ratio, as illustrated in Fig. 26. We can then immediately see how much greater or weaker the combined effects are at a certain dose ratio of A and B. If we want to analyze the data with respect to dose-additive or independent combinations, we have to realize that the relationship between additivity and independence in fixed-dose ratio combinations also depends on the steepness of the DRCs – analogously to that seen in fixed-dose studies (see Fig. 17). This implies that dose-additive combinations represent different degrees of magnitude of increased effects, i.e., small increases in case of flat DRCs and pronounced increases with very steep DRCs.

4.2.3 Multiple combinations

In principle, *dose-additive* combined effects of more than two agents can be calculated in analogy to the procedure illustrated in Fig. 24a. The only difference to the scheme in Fig. 24a is that we have to determine and add the equieffective doses of the other compounds, rather than one equieffective dose, to the selected dose of A. As with two-agent combinations, the construction of an additive DRC is only meaningful, from a mechanistic point of view, if all the compounds share the same slope and maximum. However, even under these suppositions, a *site-directed analysis* apppears of doubtful significance because complicated interactions

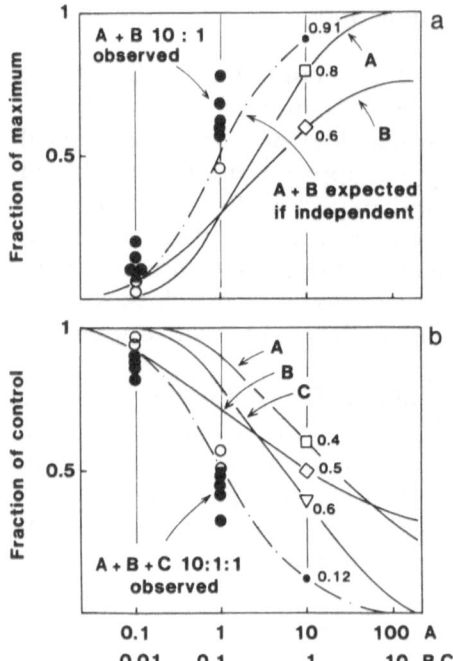

Fig. 26. DRCs of A and B (**a**) and A, B, C (**b**) alone and in fixed-ratio combinations with separate dose scales for A, B, C. The DRC of combined effects represent independent DRCs, their calculation is exemplified at the dose of 10 A. Open and closed circles at the doses of A = 0.1 and 1 indicate individual "observed" values of fraction of maximum (**a**) and of control (**b**), respectively, of the mixtures

among the components may occur. In the interpretation of results we have to consider, also with respect to independence, that the combined effects seen represent the overall combined response of the agents under study.

Independent DRCs can also be determined in multiple combinations (Fig. 26b), in analogy to two-agent combinations (Fig. 26a). The calculation is quite simple if experimental results are expressed in terms of fraction of control, illustrated in Fig. 26b. Here, an example is given for an independent action of A + B + C = $0.4 \times 0.5 \times 0.6 = 0.12$, calculated according to Eq. (6). When results of fixed-ratio experiments are expressed differently, we could possibly convert the values for effect E to $(1 - E)$ and use Eq. (5) for calculation of independent effects.

4.3 Tips and hints

When an experiment is planned in which the effects of A shall be studied in the presence of B *and* if one is interested in the site(s) of action, the dose of B and its effect should be chosen to give the best possible separation of the theoretical curves for additivity and independence. This can be checked by calculation of the theoretical DRCs for a given effect of B before an experiment, we only need to know the shape of the DRC of A, its slope and E_{max}. These points have been considered recently (Pöch et al. 1990a, b).

5 Evaluation by time-course studies

There are many areas where the time course is studied in addition to or instead of DRCs. Hence, a considerable number of publications on combined effects shows the time course of A and of B alone and combined, e.g., in biochemical studies and in experiments with growth of bacteria or tumor cells.

It is simple to evaluate the results of time-course studies with respect to independent actions, described above (Chap. 2.3), provided that the maximum possible response is known. Basically, A and B are tested in a certain dose alone and combined, e.g., 1 A, 1 B, and 1 A + 1 B, i.e., as a 1:1 mixture. The response of 1 A + 1 B can then be compared with the calculated effects of independently acting compounds.

The combined response to A and B can also be compared with the expected additive response, although this is not quite as simple. Such a comparison, however, is only important for a site-directed analysis where a dose-response study is not possible or where a time-course study should complement a dose-response study. Under the presupposition that 1 A and 1 B are (roughly) equieffective, 0.5 A and 0.5 B together should yield the same response as obtained with 1 A and 1 B alone, if they interact additively.

As with mixtures at a fixed dose-ratio in dose-response studies, most investigators are interested in the magnitude of the time course of combined response rather than in the interaction with respect to the site of action of the drugs studied in combination. For this purpose, it is not necessary to compare the combined effects with either independent or dose-additive combinations; the combined effect in a time course could also be expressed by a combined-effect graph, in analogy to DRCs (Fig. 3 and Chap. 7). However, we might be interested to express the combined effect in the *time course* with the one expected for independent actions, i.e., with situations where A does not affect the action of B, and vice versa.

5.1 Time course of independent actions

The calculation of independent actions of A and B was described earlier in Chap. 2, and applied to DRCs in Chap. 4. We can easily calculate an independent interaction in time-course studies, as outlined schematically in Fig. 27 for inhibitory effects of 1 A and 1 B. There, the time course of 1 A and 1 B alone and the calculated time course for an independent action of 1 A and 1 B combined is shown. The calculation of independent effects is

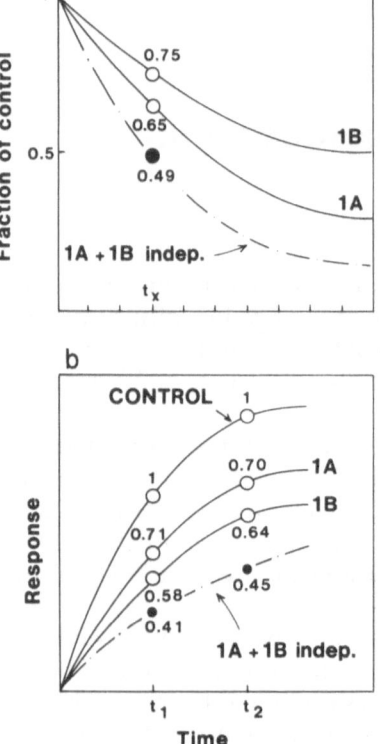

Fig. 27. Schematic time course of two drugs tested alone and their expected independent combination, expressed as fraction of control (**a**) and as fraction of maximum response (**b**) under control conditions and in the presence of drugs. The calculation of 1A and 1B acting independently in combination by Eq. (4) (Chap. 2.3.1) is shown at the time t_x (**a**) and at t_1 and t_2 in (**b**)

shown for the time t_x in Fig. 27a. If 1B reduces the control value 1 to 0.75, and 1A to 0.65, then the combination is expected to decrease the control value to $0.75 \times 0.65 = 0.49$ in independent interaction, according to Eq. (4).

Quite recently, the approach of Fig. 27b was used for the evaluation of nonsteroidal antiinflammatory agents in two rat inflammatory models (Hirschelmann et al. 1991); the new agent CGP 28238 apparently interacted by independent mechanisms with the glucocorticoid, dexamethasone, i.e., the time course of the combined effect was not significantly different from an independent time course ($p = 0.23$ and $p = 0.73$, respectively).

If the control is not (expressed as) a constant, then the inhibitory effects of 1A and 1B have to be related to each control value, as illustrated in Fig. 27b. At t_1 and at t_2, the calculated independent response in this example is 0.41 ($= 0.71 \times 0.58$) and 0.45 ($= 0.70 \times 0.64$), respectively.

5.2 Dose-additive time course

The construction of an additive time course for the combined effects of 1A and 1B would require the knowledge of the dose-response relationship of the time course, e.g., at the time the respective maximum of effect is

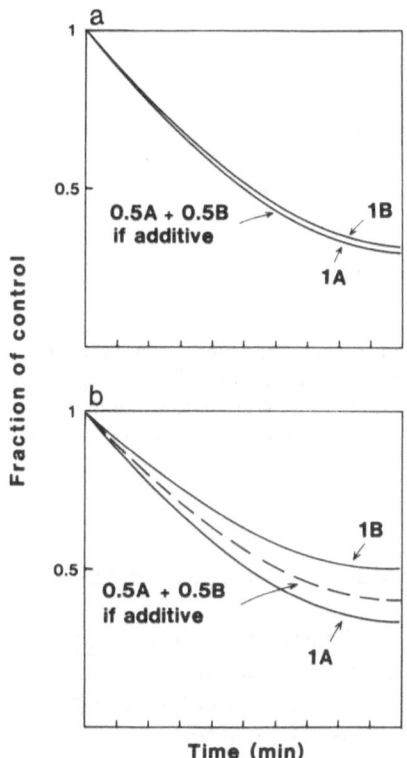

Fig. 28. Schematic time course of two drugs, analogous to Fig. 27, however, with the expected dose-additive time course. **a** 1 A and 1 B are practically equieffective, hence the additive response to 0.5 A + 0.5 B equals either 1 A and 1 B; **b** 1 A corresponds to 2 B in DRCs with slope 1, hence the additive response to 0.5 A + 0.5 B is different from 1 A and 1 B, but greater than 1 B and smaller than 1 A

reached. However, instead of calculating the effect of 2 A for equieffective doses of 1 A and 1 B, we can assume that fractions of equieffective doses amounting to 1 should yield the same response as 1 A and 1 B, respectively. This consideration is illustrated by the schematic graph in Fig. 28 a.

There is, however, a practical drawback. It would not be possible, in general, to test doses of A and B which alone are strictly equieffective, even if pretested for an equieffective response. In this case, we can only detect larger deviations from additivity as effects exceeding the time course of the more effective compound, e.g., 1 A in Fig. 28 b, although the expected time course of an additive combination will always be between the curves of A and B. However, if we know the dose-response relationship, i.e., the slope of the DRC of the more effective agent, we can calculate a theoretical time course of an additive interaction. This situation is shown in Fig. 28 b, assuming a slope of 1 of the DRCs of A and of B.

5.3 Statistical comparison of observed and expected time course

In time-course studies we can statistically compare the observed with the expected time course of dose-additive or independent interactions, analogously to dose-response studies with mixtures, described above; the com-

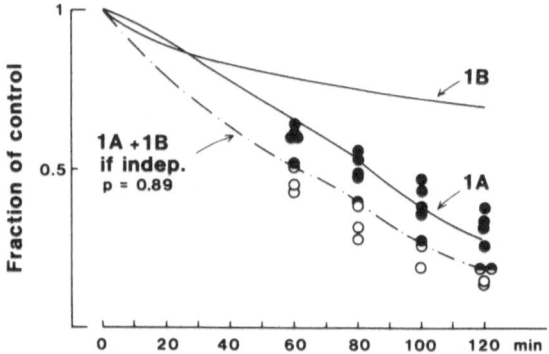

Fig. 29. Illustration of statistical comparison of observed and expected time course. The time course of 1 A and 1 B is shown by the lines which connect median values, shown in Fig. 30 b. Observed individual values of 1 A + 1 B are shown (○ values below, ● values above independent time-course line, broken dotted) only from 60 min on for graphical clarity. Calculation of χ^2 goodness-of-fit also used values at 2, 5, 10, 20, and 30 min (p = 0.89)

parison is done at time points rather than at doses for dose-additive as well as for independent interactions.

Figure 29 illustrates this procedure on the basis of an actual experiment, where the values of individual experiments are shown at 60 min and later, for graphical reasons, only. The expected response in this example is the independent time course; the actual response is a graded one.

Examples: Comparison with independence
Figure 30 illustrates the evaluation of time-course studies with respect to independent interactions of smooth muscle relaxants. In the upper panel, the combined actions are not significantly different from the expected time course of independently interacting drugs, in the lower panel, the combined effects significantly exceed independent effects. Figure 30a and b shows the time course of a β-adrenergic and a calcium antagonistic drug, salbutamol and nifedipine. It can be seen that the combined effects of these drugs correspond to independent interactions, regardless of the magnitude of effects in the individual time course. In Fig. 30a, salbutamol was the more effective drug, in Fig. 30 b, nifedipine was much more effective than salbutamol. This fact may indicate that the results reflect truely independent interactions, not likely to be caused by coincidence.

In the lower panel of Fig. 30, the combined effects of the β-adrenergic agent, isoprenaline, were markedly enhanced by the phosphodiesterase (PDE)-inhibitors, theophylline (Fig. 30c) and papaverine (Fig. 30b), resulting in a combined response exceeding the expected time course of independently interacting drugs. These results are in line with results of dose-response studies (Pöch et al. 1990a) and are explained by the sequential interaction of drugs, which lead to an enhanced production of cAMP and inhibition of its breakdown (see Pöch et al. 1990a).

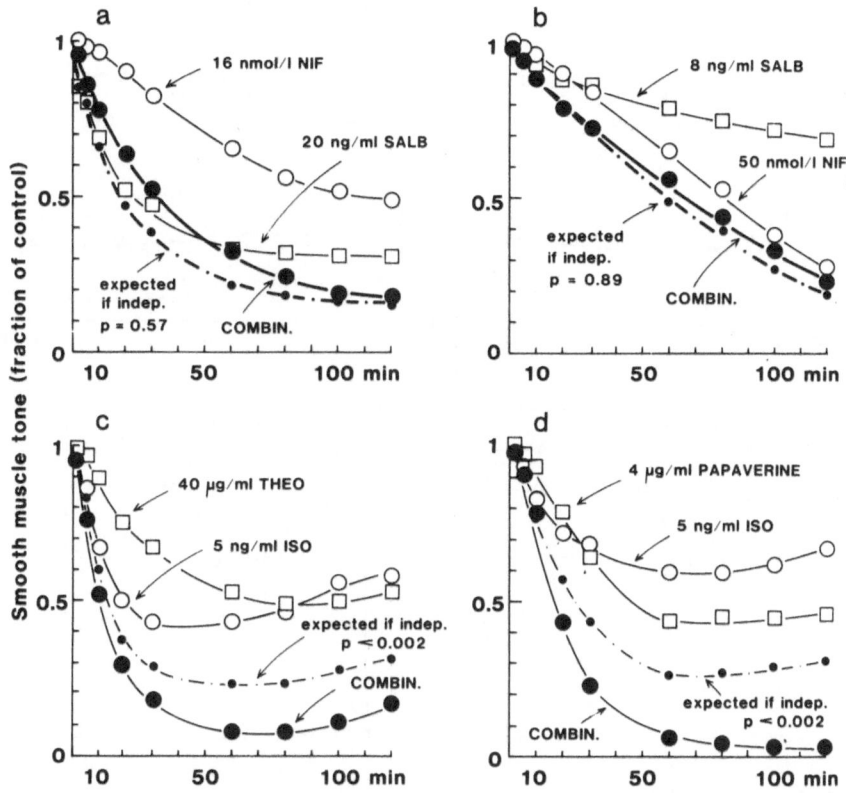

Fig. 30. Experimental time-course studies with drug combinations (*COMBIN.*). **a, b** Combinations showing no significant differences against an expected independent time course. **c, d** Combinations significantly exceeding independent effects in the time course. The drugs used were: nifedipine (*NIF*), salbutamol (*SALB*), theophylline (*THEO*), isoprenaline (*ISO*), and papaverine, applied in the indicated concentrations. Median values from 7–13 strips of bovine tracheal muscle (**a, b**), and of bovine coronary muscle (**c, d**). Statistical comparison of combined effects with independent effects by χ^2 goodness-of-fit statistics

Examples: Comparison with additivity

Figure 31 presents examples for the comparison of observed time course with the one expected for dose-additive combinations. In Fig. 31a the time course of the combination of nitroprusside-Na and nicorandil in half the doses shows a significantly greater relaxing effect than either of these compounds in "full" doses. The results can thus be interpreted as an overadditive interaction. In the presence of 1 μmol/l of the K-channel blocker glibenclamide the analogous experiment reveals an interaction which is not significantly different from additive. The contrasting results of Fig. 31a and b have been explained by Holzmann et al. (1992) by the circumstance that under the conditions in Fig. 31a nicorandil has two components of action, one of which shared with nitroprusside-Na. After block-

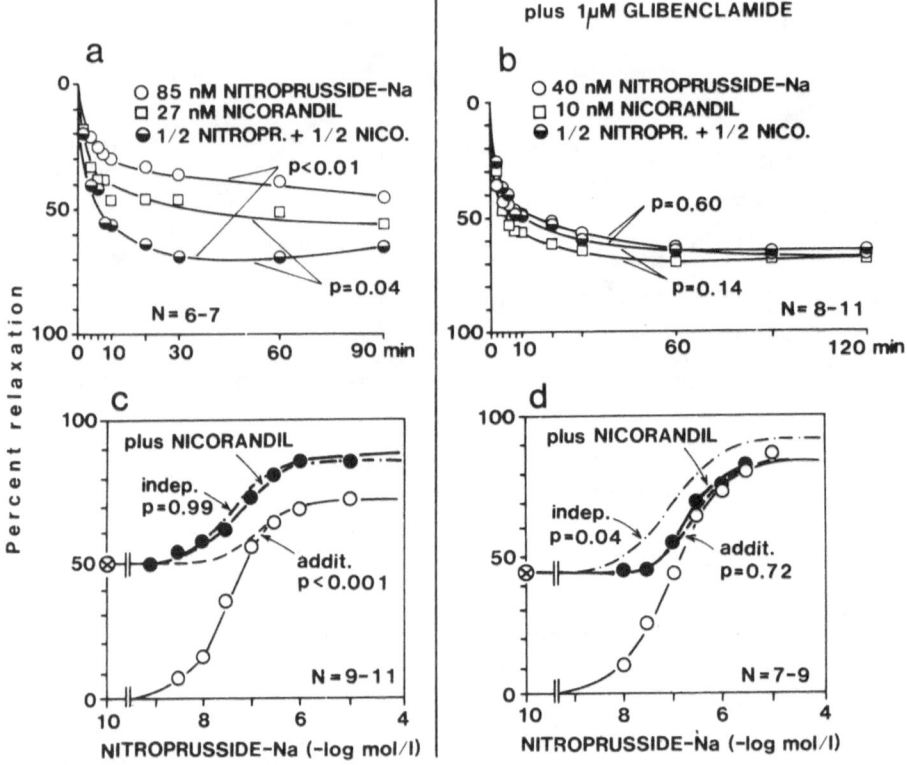

Fig. 31. Experimental time-course studies (**a, b**) and DRCs (**c, d**) with nitroprusside-Na and nicorandil in isolated bovine coronary arteries. **a, b** ○ □ Relaxant effects with full doses of compounds, ◐ effects with half the respective doses combined. **c, d** DRCs of nitroprusside in the absence (○) and presence (●) of 50 μmol/l nicorandil, analogous to Fig. 22; the slope values of nitroprusside in the absence of nicorandil were 1.1 and 1.2, respectively. **a, c** Results in the absence, **b, d** in the presence of glibenclamide. Statistical comparison of experimental with calculated additive and independent response by χ^2 goodness-of-fit test

ade by glibenclamide of the second component, not shared by nitro-prusside-Na, nicorandil behaves simply like nitroprusside-Na.

5.4 Time-course studies with log response scale

As with DRCs, the time course of effects is not infrequently expressed by changes in response on a log scale, e.g., with chemotherapeutic agents. Figure 32 schematically shows such a time course for the growth of bacteria (a) and an experimental example (b). In analogy with DRCs expressed with effects on a log scale (Fig. 8), the same relative effect is seen by the same distance in time-course studies. In Fig. 32a, at the time 1, A produces a 90% decrease in growth in the absence and in the presence of B. Hence, in the presence of B, the combined response at t = 1 represents the response expected for independently interacting agents.

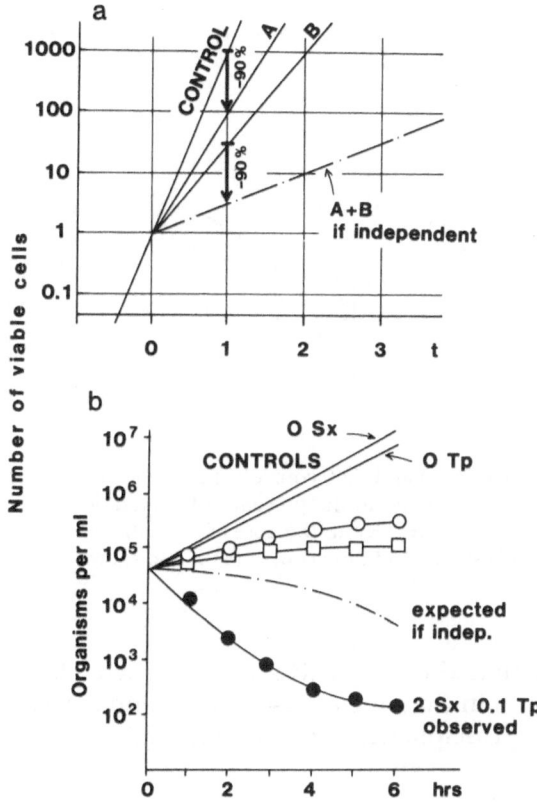

Fig. 32. Schematic (**a**) and experimental (**b**) time course (linear scale) of cell viability (log scale) of two drugs, tested alone and expected in combination if they had interacted independently. **a** The independent inhibition of viable cell number is 90% with A at t = 1, alone as well as in the presence of B. **b** The inhibitory effects of 2 µg/ml sulfamethoxazole (○), 0.1 µg/ml trimethoprim (□) alone, and in combination, observed (●) and expected if independent. Data from figures 2 and 3 of Amyes (1982) with E. coli K-12

Figure 32b shows the inhibitory effects on the number of colony forming units of 0.1 mg/l trimethoprim and 2 mg/l sulfamethoxazole alone and in combination. It is clearly evident that the combined response markedly exceeds the sum of the individual decrease produced by trimethoprim and sulfamethoxazole. This example illustrates the well known combined effect of a sequential blockage in bacterial synthesis of tetrahydrofolic acid (Black 1963, Harrap and Jackson 1975).

5.5 Comparison of time-course studies and dose-response studies

The relationship between time-course studies and DRCs is easily understood if we consider that the time course shows the development of effects, and DRCs (usually) show the effects which are finally reached, i.e., at the steady state. This relationship is illustrated in Fig. 33 for the effects of the doses 1 of A and of B, and for their effects of these doses combined. It is clearly seen that, e.g., the effect of 1 A amounts to 50% in the DRC (Fig. 33a), an effect that represents the maximum effect in the time course (Fig. 33b). It can also be seen that in this example the combined effects exceed the calculated effects for an independent action of A and B in the DRC as well as in the time

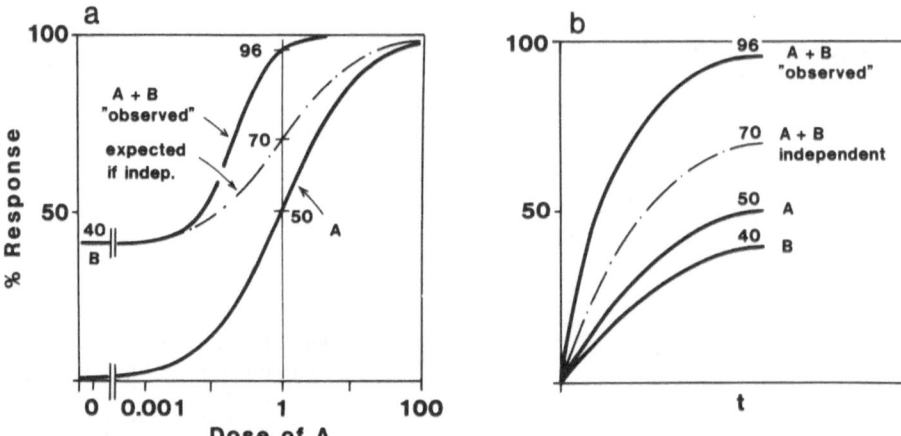

Fig. 33. Schematic comparison between DRCs (**a**) and time course (**b**) of two agents, A and B, alone and in combination. **a** DRCs of A alone and in the presence of a fixed dose of B, **b** corresponding time course of A, B, and A + B. A represents the dose of 1 A, B the fixed dose of B. The calculated independent DRC (**a**) and time course (**b**) is also shown

course. Hence, we can conclude that whenever DRCs are constructed from points representing the maximum in the time course, a DRC and a corresponding time course should in principle show the same result.

Examples

This is indeed observed experimentally, as demonstrated by examples. In Fig. 30a and b the combined effects of salbutamol and nifedipine were not significantly different from independent combinations. This result was also obtained in dose-response experiments performed under comparable conditions (observed combined effects vs. independent effects: $p = 0.11$, not shown). Greater than independent interactions between isoprenaline and papaverine in time-course studies (Fig. 30d) are reflected by significantly greater than independent effects in DRCs (Pöch et al. 1990a). Similar results have been observed with isoprenaline and theophylline in time-course studies (Fig. 30c) as well as in DRCs (not shown).

Figure 31 also demonstrates that, in principle, a comparison of observed with additive combinations provides the same results in time-course studies (Fig. 31a and b) as in dose-response studies (Fig. 31c and d). In the absence of glibenclamide, the combined effect of nitroprusside-Na and nicorandil was significantly overadditive (Fig. 31a and c) but was additive in the presence of glibenclamide (Fig. 31b and d).

5.6 Conclusions

From the theoretical considerations and from practical experience we can conclude that combined effects in time-course studies can also be compared with additivity and independence, and that these results reflect results expressed by DRCs. A comparison with independent effects can easily be carried out in time-course studies but a comparison with additivity requires roughly equieffective doses of the parent compounds. In other words, the greater the differences between A and B, the more difficult it may be to detect over- or underadditive combinations in time-course studies. However, a comparison with additivity appears of less importance, in general, than a comparison with independence, as already discussed in the foregoing chapters.

6 Single-dose and other studies

6.1 Single-dose studies

In certain situations neither DRCs nor time-course studies are obtained, yet combined effects are evaluated. The most simple approach is to study the effects of single doses of A and B (C ...) alone and combined, in full or reduced doses, e.g., where DRC- and time-course studies are not possible or not appropriate. Hence, single-dose studies are done in all disciplines, e.g., in orienting experiments, or in clinical pharmacology and in toxicology.

Regardless of whether or not combined effect experiments are aimed at a comparison with *independent* effects or with effects expected for *dose-additive* combinations, the questions behind single dose studies are often the same basic ones asked in DRC- and time-course studies. These are mainly, whether we can get a greater effect in combination with the same doses or whether we can obtain the same effect with reduced doses (of A and B) in combination.

The effects of single doses of A and B (C ...) in combination are quite often compared with the sum of the individual effects, i.e., with *effect-additive* combinations (e.g., from a pragmatic point of view: Arcos et al. 1988). Some authors consider combined effects greater and smaller than effect-additive as synergism and antagonism, respectively (e.g., Heagle and Johnston 1979, Müller and Von Felten 1986).

6.1.1 Comparison with effect-additive

Müller and Von Felten (1986) studied the effects of a number of drugs which by different mechanisms inhibit platelet aggregation, in single doses alone and combined, and compared the effects in combination with the sum of the individual effects. As is evident from Table 5, the effects of 11 combinations exceeded the sum of the effects ($E_A + E_B$), whereas 4 combinations resulted in smaller effects than effect-additive.

Hanna and Roth (1979) studied the antagonistic effects of ephedrine or salbutamol and theophylline in combination, with a few doses of the drugs, against 5-HT- and acetylcholine-induced airway spasm. This study was aimed to test the proposal that "components of combination bronchodilator compounds interact in a synergistic manner to relax airway smooth muscle contraction" (Hanna and Roth 1979). The results showed that

Table 5. Single-dose studies with drug A, B (singly), and in combination (A + B)

A	B	% inhibition of platelet aggregation				
		E_A	E_B	obs. E_{A+B}	$\Sigma E_A + E_B$	calc. indep.
Verapamil	Diclofenac	22	24	56	46	41
	Ceftriaxone	31	38	57	69	57
	Propranolol	24	30	43[a]	54	47
	Theophylline	20	15	42	35	32
	Dipyridamol	23	30	54	53	46
Dipyridamol	Ceftriaxone	31	30	76	61	52
	Propranolol	31	32	65	63	54
	Theophylline	26	20	47	46	41
	Diclofenac	30	31	56	61	52
Diclofenac	Ceftriaxone	28	33	62	61	52
	Propanolol	28	32	49[a]	60	51
	Theophylline	27	12	41	39	36
Theophylline	Ceftriaxone	14	39	61	53	48
	Propanolol	19	15	55	34	31
Ceftriaxone	Propanolol	16	34	68	50	45

Data from Müller and Von Felten (1986)
Observed effects of A and B in combination (*obs. E_{A+B}*) are compared with the sum of individual effects ($\Sigma E_A + E_B$) and calculated independent interactions (*calc. indep.*)
[a] Less than independent

"combinations of either ephedrine and theophylline or salbutamol and theophylline interacted in vitro in an additive fashion" (Hanna and Roth 1979). The authors further interpreted their data as in disagreement with others, who claimed a synergistic interaction.

Another example is the foliar injury of soybeans exposed to ozone (O_3) and sulfur dioxide (SO_2), published by Heagle and Johnston (1979), in which the effects of the mixture of O_3 and SO_2 were compared with the sum of effects. These data are shown in Table 6. Six combinations gave effects greater than effect-additive ("observed" vs. "sum"), 2 mixtures represent effect-additive combinations, and 7 mixtures exhibited effects smaller than the sum of effects. In 4 of the latter, the calculated sum of individual effects is greater than 100%, with 100% representing the worst possible frequency of foliar injury.

A data base on combined effects of carcinogens lists 698 combinations of which 403 (58%) were greater, 16 equal to (2%), and 279 (40%) less than effect-additive, termed synergism, additive, and antagonism (Arcos et al. 1988). In the introduction, the editors indicated their definitions of synergism, additivity and antagonism as "very broad" (Arcos et al. 1988).

Another interesting example is provided by the effect of cyclosporine A and mizoribine on *graft survival*, where the effect is time prolongation. Hayashi et al. (1990) reported that in different allograft models these agents in combination led to a pronounced increase in survival time de-

Table 6. Effects of O_3 and SO_2 alone and in combination

% foliar injury of soybeans

O_3 (ppm)			SO_2 (ppm)			$O_3 + SO_2$		
0.25	0.5	1	0.5	1	1.5	obs.	calculated	
							sum	indep.
2				4		14	6	6
	21			4		36	25	24
		67		4		81	71	68
2					93	51	95	93
	17				93	77	> 100 !	94
		57			93	80	> 100 !	97
6			0			4	10	6
6				4		10	10	10
6					63	59	69	65
	56		0			56	56	56
	56			2		61	58	57
	56				75	89	> 100 !	89
6				9		26	15	14
	40			9		56	49	45
6					89	75	95	90
	40				89	77	> 100 !	93

Data from Heagle and Johnston (1979)
Observed combined effects (*obs.*) are compared with calculated effect-additive (*sum*) and independent combinations (*indep.*)

spite the fact that 2.5 mg/kg/day mizoribine alone was without effect. For example, the survival time of the kidney model due to 10 mg/kg/day cyclosporine A was 12.2 ± 2.9 (SD) days compared with the nontreated control (8 ± 1.4) but the survival with the mixture in these doses amounted to 93.6 ± 61.1 days (n = 5). Had both agents prolonged survival time, an effect-additive evaluation could have been performed.

6.1.2 Comparison with independence

Some authors directly compare combined effects with those calculated for independent interactions (e.g., Yamaji et al. 1990, Elashoff et al. 1987; termed "additive" by Colby 1967). Quite often, however, the sum of effects is considered for comparison with observed effects in combination, as described above. The latter can be interpreted with respect to independent interactions although not without reservations.

In Chap. 2 the relationship between effect-additive and independent interactions has been described (Chap. 2.4.2; see also Chap. 3.3.2, Fig. 18). It was discussed that only greater effects than effect-additive allow the conclusion that they represent effects greater than those of independent combinations, regardless of whether or not the response is expressed with respect to a possible maximum. Hence, where effects cannot be expressed with respect to the maximum possible response, combined effects exceeding the sum of the individual effects must also represent greater effects than independent (Fig. 18). On the other hand, where the effects can be expressed in terms of percent maximum response, we prefer to calculate independent effects "accurately" (see Chap. 2.3.1), e.g., for the data of Tables 5 and 6.

In Table 5, with the exception of two, all combinations showed effects equal to or greater than independent. The two combinations which exhibited effects below those of independent combinations, showed slightly weaker effects than expected for independently interacting drugs (43 vs. 47%, and 49 vs. 51%). In summary, we may conclude that the overall combined effects, listed in Table 5, were equal to or greater than independent, suggesting no interaction in some cases, and some kind of (sequential) synergistic interaction between drugs. Different mechanisms referred to by the authors (Müller and Von Felten 1986), by which the calcium concentration of the platelets were decreased by drugs, may explain these interactions. These to some extent, or in some cases, may operate in a dependent or functional cooperative manner. The results of these studies with platelets have their counterpart in results with smooth muscle relaxants acting at different sites (e.g., in Fig. 30 c and d) (see also Chap. 8, and Pöch et al. 1990a).

Table 6 shows that the toxic effects of the air pollutants on soybeans were greater than independent in 6 combinations, equal to independent in 3, and less than independent in another 6 combinations. So, we may conclude that the overall effect in combination was comparable to independent interactions, rather than to "variable responses ... to mixtures" (Heagle and Johnston 1979).

The examples in Tables 5 and 6 show that when A and B are tested in the same doses alone as well as in combination, a comparison of observed with independent interactions is possible. The appropriate equations can be chosen from those given in Chap. 2.3.1.

Let us consider other examples here to further illustrate this important practical approach (additional examples will be given in Chap. 8). They often show that combined effects correspond to independent effects or exceed them.

Differentiation and inhibition of cell growth
Yamaji et al. (1990) studied the combined effects of tiazofurin and retinoic acid on differentiation and inhibition of colony formation of HL-60 leukemic cells because "it became of interest to test the hypothesis that tiazofurin which down-regulates the ras oncogene ... and retinoic acid, which down-regulates the myc oncogene ... might be able to interact synergistically."

Differentiation was measured as percent nitroblue tetrazolium positive cells and showed an increase from the control value of 3 ± 2 to 12 ± 4 by retinoic acid and to 29 ± 4 by tiazofurin alone. Combined these compounds led to an increase to 68 ± 6, which was far, and statistically significantly, in excess of the predicted value (38%) for independent interaction (Yamaji et al. 1990: table II). A similar interaction was observed in this study on colony formation, which was reduced in combination from 100% to 44%, again by far and significantly exceeding the predicted value of 77% for an independent combination (Yamaji et al. 1990: table III).

Depressant effects on myocardial contractility
Caplan and Su (1986) have investigated the effects of the anesthetic, halothane, and the calcium-antagonist, verapamil. They found that a single dose of verapamil alone decreased peak developed tension to $85 \pm 2\%$, halothane alone to $44 \pm 2\%$, and both drugs decreased tension to $32 \pm 2\%$ of control. The authors interpreted the combined effects to be "almost identical to the effect that would be predicted for the sum of the depressant effects of each drug alone." A comparison with independent effects appears more appropriate and meaningful. It gives an expected effect for an independent combination of 37% of the control. Hence, the combined effect of halothane and verapamil, described by Caplan and Su (1986) can be interpreted as being more pronounced than independent.

Recurrent bleeding in ulcer patients
Recurrent bleeding is one important risk in patients with gastroduodenal ulcer. Therefore, two drugs were tested in single doses in combination and compared with the effects of the drugs applied singly in a double-blind multicentre trial (Londong et al. 1982). The results showed successful prophylaxis in 18 out of 31 (58%) patients with the anticholinergic drug, pirenzepine, in 26 out of 35 (74%) with the H_2-antagonist, cimetidine. In combination, the rate of success was 91% (31 out of 35 patients), comparable to the calculated effect of independently interacting agents (90%).

Antisecretory effects
Combinations of the anticholinergic drug, pirenzepine, with H_2-receptor antagonists, showed significantly greater inhibition of gastric acid secretion in healthy volunteers than expected for independent interactions (Pöch and Londong 1985). Pirenzepine caused $60 \pm 4.0\%$, cimetidine $61 \pm 4.6\%$, and the combination $90 \pm 0.8\%$ ($n = 8$). The calculated independent effect in combination was $83 \pm 3.1\%$.

If we accept that combined effects which exceed the effects of independently acting compounds or factors, can be considered as evidence for potentiation (see Chap. 3), at least where the components of a combination cannot *act* alike, a comparison with independence can also be interpreted with respect to potentiation. The above example can thus be interpreted to indicate a potentiated response in combination, because it exceeded the calculated effect of independently interacting agents. Pöch and Londong (1985) interpreted it somewhat differently, i.e., as potentiation because it was believed to represent an overadditive combination.

Antitumor activity

Kim et al. (1990) reported that interleukin-2 (IL-2) and tamoxifen reduced experimental metastases by 66% and 30%, respectively. IL-2 and tamoxifen combined reduced metastases by 95%, significantly better than did IL-2 ($p < 0.02$) or tamoxifen ($p < 0.0003$) alone. The expected pharmacodynamic effect would have been 76% for independent action of these agents.

As a general guideline, we should keep in mind that experimental results of single-dose studies may not always be confirmed by dose-response or time-course studies. For instance, deviations from additivity may be overlooked if they occur at other doses.

6.1.3 Comparison with additivity

When A and B in single-dose studies are tested in "full" doses alone and in reduced doses in combination, a comparison with additivity is possible but not appropriate for a site-directed analysis, where we should preferably perform dose-response studies.

Single-dose studies, aimed at doses for a certain effect, can be carried out by testing A and B alone and by testing fractions of equieffective doses (concentrations) in combination. If 1A and 1B indicate equieffective doses, than 0.5 A + 0.5 B is expected to give an additive response, i.e., the same response as 1A or 1B. The same applies to other fractions of equieffective doses in combinations which sum up to 1.

There are situations where such extensive studies are not feasible or are not aimed at, e.g., clinical studies, certain toxicological investigations or studies with chemotherapeutic agents. For instance, we might be interested to see whether the minimum inhibitory concentration (MIC) of antibacterial agents can be reduced in combination without loosing the required effect.

Published single-dose studies with reduced doses of the components in combination are scarce. Therefore, two examples are discussed here. In the first study, the antiarrhythmic effects of sotalol and aprindine in patients were evaluated (Stroobandt et al. 1987) in "full" doses of 1.5 mg/kg of sotalol, 2 mg/kg of aprindine, and in half doses of both. The compounds significantly reduced the frequency of ventricular arrhythmias by 41 and 51%, respectively, when applied alone, and by 72% in combination. Hence, the combined effect appeared greater than expected for an additive combination (41 to 51%) – but the differences between the different treatments were not significant.

The second example, is another clinical study (Dal Monte et al. 1985), in which pirenzepine (150 mg) and cimetidine (1 g) were given daily for ulcer healing alone and in about half the doses combined (75 mg plus 0.4 g). The frequency of ulcer healing was 35 and 41% in single-drug therapy, and 84% in combination therapy, significantly higher than with pirenzepine or cimetidine alone.

6.1.4 Tips and hints

In analyzing combined effects of single-dose studies the same tips and hints can be given as with DRC- and time-course studies, i.e., first to check how the terms "additive" and "independent" are used in the literaure. It appears that many discrepancies are due to different terminology.

When combined effects are expressed with respect to the *sum* of individual effects and when they significantly exceed it, greater than independent effects by necessity can be assumed. Lower than effect-additive responses do not allow any conclusion with respect to *independent* interaction but require the calculation of independent combinations for comparison. This calculation is described in Chap. 2.3.1, including consideration of background "effects". Once again, the attention of the reader is directed to the sound postulate that greater than independent effects indicate potentiation (see Chap. 3).

A comparison with *additivity* requires testing of "full" doses of A and B alone and reduced doses of A and B in combination, e.g., half the doses. Testing for the same binding site should not be done with single doses, except for orientation, since deviations from additive might appear in other areas of the DRCs.

6.2 Other studies

Combined effects are also investigated and expressed in different ways than described so far in this book. For example, in the analysis of risk or by indicating the survival time in combination. In some of these studies it may be possible to express the results with respect to frequency or with respect to the time course. For instance, risk studies are often derived from frequency studies, which in turn, can be evaluated with respect to independent interactions (Pöch 1991 b). Also, survival time is derived from the time course of survival, and hence allows an analysis of combined effects with respect to independence, when the time course for the singly applied compounds or factors is also studied.

Other ways to present combined effects in any field may be checked whether they can be tested and evaluated as described above, preferably with respect to the dose response or the time course of events. Alternatively, combined effects might be expressed or evaluated as single-dose studies. In any case, we need to know the combined effects of chemical compounds or physical factors as well as the effects which they produce when "acting" alone.

7 Combined-effect graph and other graphs

Combined effects can be displayed graphically in different ways, e.g., three-dimensionally by response-surface diagrams (e.g., Greco 1987, Prichard and Shipman 1990, Sühnel 1992), in principle invented by Loewe (1953), or two-dimensionally by isobolograms, DRCs or histograms. This chapter deals with a new graph, termed combined-effect graph, and a graphical comparison of observed with calculated independent effects. Also, reference is given to the Yonetani–Theorell plot.

7.1 The construction of the combined-effect graph

The principle of a graphical plot of enhancement or diminution of effects has been outlined in Chap. 1, illustrated in Fig. 3. Here, this plot is described and discussed from a practical point of view, with respect to the methodological procedure and its application to experimental results.

Since this plot shows the enhancement of effects of A by B as a function of the effect of B, it is especially useful for demonstrating enhancement in experiments with multiple DRCs, examples of which have been selected for Fig. 34. This figure displays DRCs expressed either by the percent of maximum (Fig. 34a) or by the percent of control response (Fig. 34c). The combined-effect graphs derived therefrom are described for these two situations. In both cases, a rather low response level of A was selected, from which the enhancement of A by B was determined, in this example 10% effect and 90% of control, respectively.

7.1.1 Percent of maximum response

Figure 34a and b illustrates the enhancement of *effects* of A, expressed as percent of maximum, in the DRCs and in the combined-effect plot, respectively.

In the DRCs, an enhancement of effects of A (APTS) by B (ACV) from 10% to 27, 53, 58, and 88% is seen on the vertical line, indicated on the right y-axis of Fig. 34a. This enhancement occurred at doses of B (ACV) which alone caused 8, 26, 32, and 73% effect, indicated on the left y-axis of Fig. 34a.

The combined-effect graph in Fig. 34b shows the increases of A by B on the y-axis as a function of the effect of B alone on the x-axis. It is evident that the arrows of enhancement increase with increasing effects of B, exceeding a diagonal line. The latter indicates the enhancement of effects expected

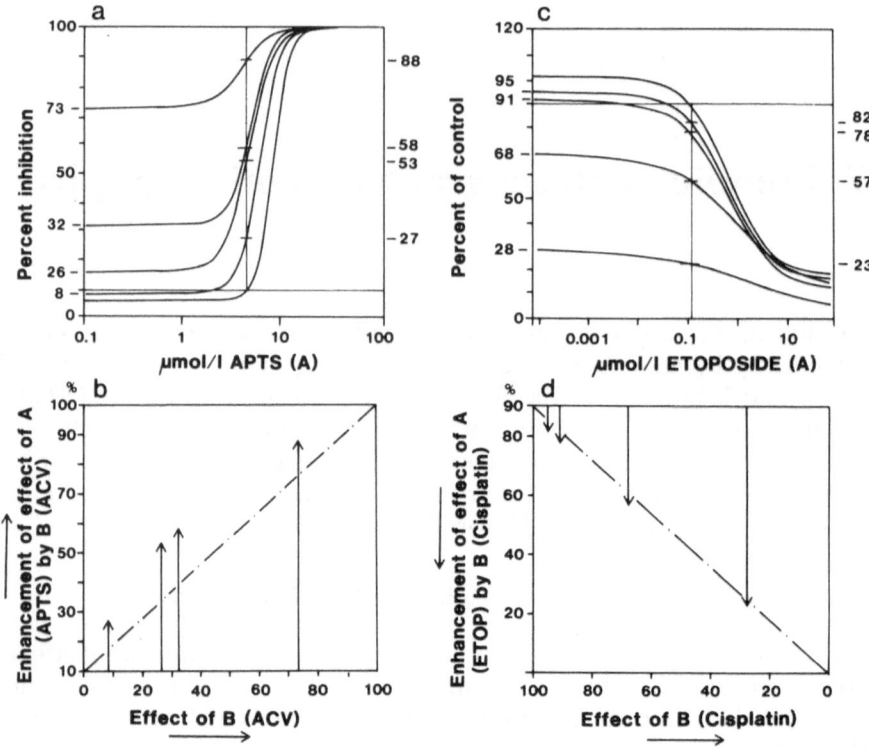

Fig. 34. Experimental DRCs (**a, c**) and combined-effect graphs (**b, d**) derived from DRCs. **a** DRCs of antiviral effects of APTS are shown in the absence and presence of various fixed concentrations of ACV, expressed as percent inhibition. **c** DRCs of antitumor effects of etoposide in the absence and presence of cisplatin, expressed as percent of control cell survival. The DRCs were constructed by ALLFIT of data published by Prichard and Shipman (1990: fig. 2) (**a**) and Tsai et al. (1989: table 2) (**c**). **a, c** The vertical lines and the scales on the right hands indicate the effects of A in the presence of B (short cross bars on the DRCs) measured at a dose where A showed a response of = 10% inhibition (**a**) and 90% of the control (**c**), respectively; **b, d** the respective enhancement of inhibition is shown by the arrows in the combined-effect plots where the diagonal lines represent expected values for independent action of the drugs. Note that the x-axes in **b** and **d** show the effects of B alone, indicated on the scales on the left hands in **a** and **c**

for independent action of A and B. For instance, at 50% effect of B and 10% effect of A, the independent effect amounts to 55%, i.e., corresponding to a fraction of increase which is 0.5. As a matter of fact, in an independent action of A and B, the fraction of increase by B determines the fraction of increase with A + B at any level of response to A (illustrated in Fig. 10).

7.1.2 Percent of control response

Analogous to Fig. 34a, the enhancement of effects of A by B is evident and illustrated by the vertical line drawn from the response level, 90% of *control response* in Fig. 34c, and indicated on its right y-axis. This enhancement of reduction occurred at doses of B which by itself caused more or less reduction, indicated on the left y-axis of Fig. 34c.

It is evident that the more reduction is elicitated by B, the more reduction in combination of A + B occurred. However, this relationship is more clearly seen when the percentage-of-control values in combination are plotted against the respective values observed with B alone in the combined-effect graph (Fig. 34d). Here, lower values than those on the diagonal line represent enhanced effects which exceed the corresponding effects of independent interactions. In this example, greater than independent effects are hardly seen.

The results in Fig. 34c and d could be replotted as percent inhibition rather than as percent of control after converting the latter values to percent-inhibition values (not shown).

7.2 Comparison of observed with independent effects

The combined-effect graph already enables a comparison of observed combined effects with independent effects, where the observed effects in combination were expressed as enhancement of effects of A, from a certain effect level, by B. From a practical point of view, we may be interested to see the relationship between observed and calculated independent effects irrespective of a certain effect level, e.g., for all actual data points.

For this purpose, the observed effects in combination are displayed on the y-axis in comparison with the respective independent effects on the x-axis, e.g., in Fig. 35b and d. For illustration, the two examples of multiple DRCs in Fig. 34 have been taken and are again shown in Fig. 35a and c, however, with observed *data points* and the respective *independent DRCs*. In Fig. 35b and d, the comparison of observed values with expected independent values are plotted, derived from the points of these DRCs.

Figure 35b shows that the points are scattered around a 1:1 relationship which may indicate an independent antiviral interaction of APTS and ACV. Figure 35d even more clearly exhibits an interaction of the chemotherapeutic agents, etoposide and cisplatin, that can be interpreted as purely independent.

Practical considerations
It is of practical interest that such a comparison between observed and expected independent effects does not require DRCs; neither does the calculation of independent effects. Hence, this plot is applicable whenever independent effects can be calculated. This also enables a simple graphical overview of various combined effects studied in a certain field, e.g., in clinical pharmacology or in toxicology, examples of which are given in Chap. 8.

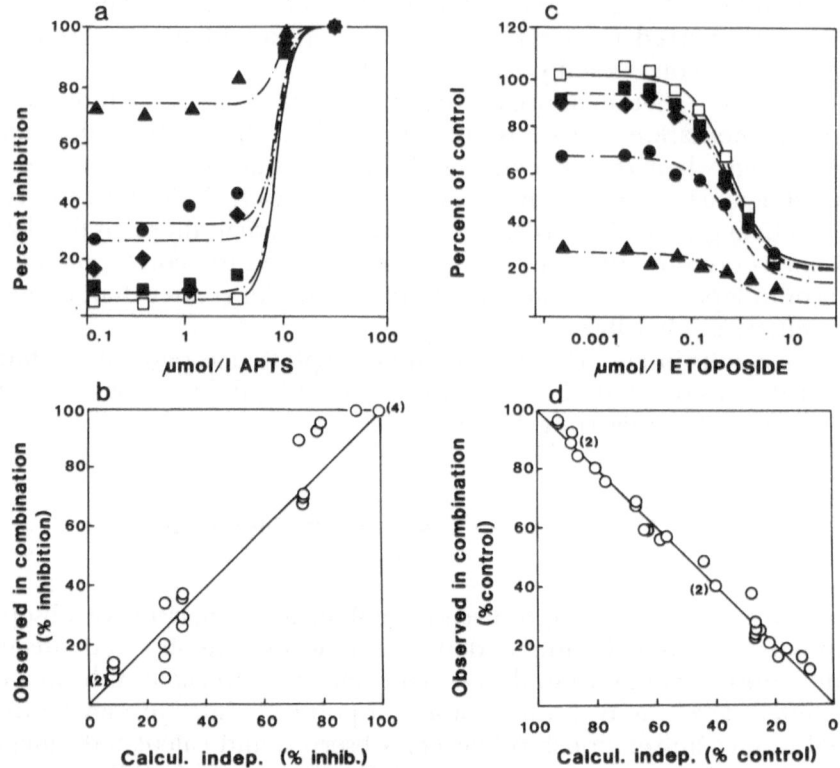

Fig. 35. Experimental DRCs (**a, c**) and graphical comparison of observed values with calculated values of independently acting agents in combination (**b, d**). **a** Experimental data points of the DRCs, shown in Fig. 34a, and **c** experimental data points of the DRCs shown in Fig. 34c rather nicely fall on the corresponding DRCs for independent effects (broken dots). **b, d** Comparison of observed values with respective calculated values of independently acting agents, from **a** and **b**, respectively. Note that the DRCs in Fig. 34 were obtained by fitting experimental points to curves, whereas the DRCs in Fig. 35 represent theoretical curves, expected for independently interacting agents

7.3 Yonetani–Theorell plot for enzyme inhibitors

Whereas the foregoing plots compare combined effects with independent effects, the plot of Yonetani and Theorell (1964) compares effects of two enzyme inhibitors with "mutually *exclusive*" and "mutually *nonexclusive*" interactions. Mutually exclusive inhibitors compete for the same molecular binding site, mutually nonexclusive bind to different sites (Wong 1975). The term "mutually exclusive" describes the competitive synergistic interaction of two inhibitors, hence a dose-additive combination. "Mutually nonexclusive" is the negation of "mutually exclusive", it means that inhibitors do not share a common molecular binding site, i.e., that they do not

act additively. There are several types of "mutually nonexclusive" interactions, *one* of which is an independent interaction, characterized by the
fact that one inhibitor does not change the dissociation constant K_i of the
other inhibitor.

In this type of graph, the concentrations of one inhibitor (A) are plotted on the x-axis vs. the reciprocal of the velocity (v) of the enzyme reaction
in the absence and in the presence of fixed concentrations of the other
inhibitor (B). Figure 6c and d shows practical examples. Parallel lines, as
in Fig. 36c indicate mutually exclusive binding, hence a competitive = additive interaction. This fact is also evident from the DRCs in Fig. 36a. The
experimental points lie on the DRCs of theoretical additivity, with which

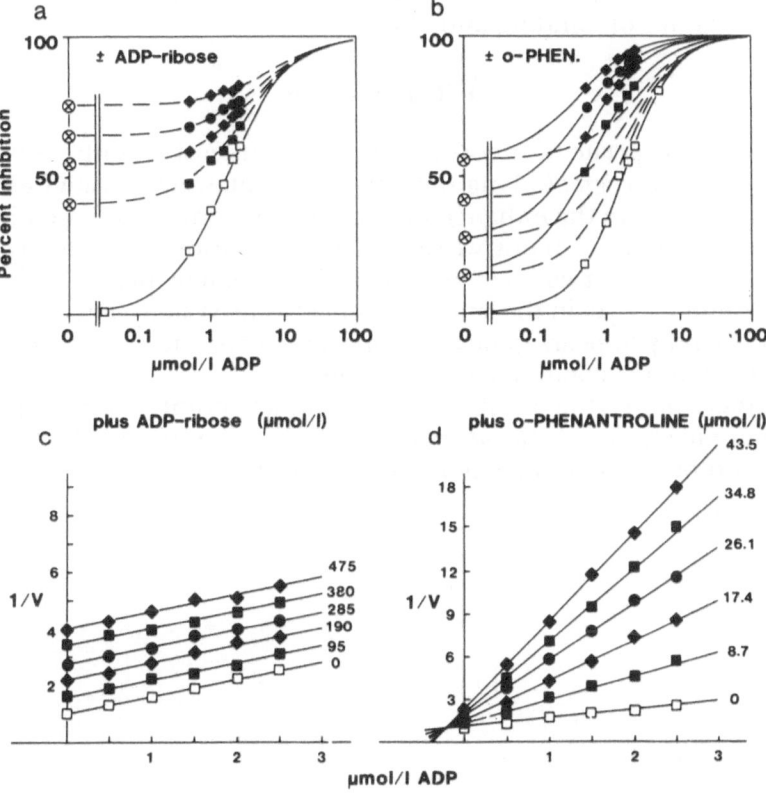

Fig. 36. Comparison of DRCs (**a, b**) and Yonetani–Theorell plots (**c, d**). Shown are the
inhibitory effects of ADP alone (□) and in the presence of the second inhibitor (closed symbols), ADP-ribose (**a, c**) and o-phenanthroline (**b, d**), respectively. The results are either
expressed as percent inhibition (DRCs) (**a, b**) or as 1/v of the enzyme reaction (**c, d**). The
closed symbols indicate the "effect" at the concentrations referred to in the lower panel. Note
that competitive interaction exhibits an additive response (broken lines) in the DRCs (**a**) and
is characterized by parallel lines in the Yonetani–Theorell plot (**c**), whereas noncompetitive
interaction shows overadditive combinations (**b**) and intersecting lines (**d**). Tabulated data
from Chou and Talalay (1981) were taken to construct DRCs and the Yonetani–Theorell plots

the experimental DRCs coincide. Therefore, the latter curves are not shown.

Non-parallel lines indicate mutually nonexlusive binding, the resulting interaction is independent *if* the lines intersect at the x-axis. In Fig. 36d, they intersect above the x-axis, and hence indicate that o-phenanthroline has lowered the K_i for ADP. This means a decrease in the ED_{50} of the DRCs, which is evident from Fig. 36b. It amounted to a dose ratio of 2.7 to 4 at low to high concentrations of o-phenanthroline.

The comparison between the Yonetani–Theorell plot and corresponding DRCs, illustrated in Fig. 36 will be dealt with in more detail in Chap. 8. The Yonetani–Theorell plot nicely illustrates whether or not inhibitors bind to the same or to different sites. It is somewhat limited to inhibitors with slopes around 1 in DRCs to obtain straight lines in the Yonetani–Theorell plot. Despite its limitation, it is based in a straight-forward manner on binding and effect.

7.4 Tips and hints

It appears appropriate to choose an effect level or level of control response for a *combined-effect graph* along the line of the following considerations. If we are interested in the enhancement *at* a certain effect level, we will base the graph on this level. If not, we can make our decision after looking at the DRCs. We will note at which effect level of A the changes produced by B are most pronounced yet in the submaximum effective range. The effect level derived thereby appears generally appropriate for the construction and interpretation of a combined-effect graph.

Further, there is the possibility to display some points in this graph, and even more likely, in the *plot of observed vs. independent effects*, either by symbols, letters or numbers attached. The reader will find some suggestions in Chap. 8.

8 Applications of the new approach and observations

This chapter describes results of practical experience with combinations of various compounds and factors in various fields and also refers to relevant publications. Many examples in the literature, which have been collected over years, and unpublished results from our laboratory will be evaluated and discussed here, mainly using the new approach, and analogous formats. These examples will show the impact of this approach and will indicate its wide, albeit not universal, applicability. Examples have also been described and discussed in Chaps. 4–7.

8.1 Biochemistry and physiology/pathophysiology

Textbooks and articles in this field provide a number of general principles and examples by which endogenous and exogenous compounds act together in a synergistic (e.g., Pöch 1983, Gibbins 1989) or antagonistic manner. For instance, allosteric activators and inhibitors of enzymes increase and decrease the rate of enzymatic reactions, respectively, by changing K_m and V_{max} (e.g., Jungermann and Möhler 1980, Pöch and Juan 1990). Hormones and second messengers, mainly by phosphorylation reaction, activate or inhibit enzymes (Cohen 1980, Shacter et al. 1984). Interesting examples are the marked increase in phosphorylase-kinase activity by calcium plus magnesium (King and Carlson 1981) or the potentiating effect of adenylate cyclase activation by forskolin (Seamon and Daly 1981) and by cholera toxin (Perkins et al. 1978). Other examples may easily be found in the literature (see also Pöch 1983).

8.1.1 Dual inhibition of enzyme activity

The results of studies with two inhibitors appear of special interest with respect to the site of action and type of interaction. They provide impressive evidence that inhibitors which bind to the same site interact competitively, whereas those which bind to different sites on an enzyme exhibit a combined effect which is distinct from competitive interaction. This topic has been described by Cleland (1970), by Wong (1975), and by others (e.g., Chou and Talalay 1981).

In a *site-directed analysis* of combined effects we are primarily interested whether or not the results of combination experiments support binding to the same molecular site, resulting in an *additive*, i.e., competitive interaction. This evaluation can be done by DRCs, as described in this book, but is also possible by the plot of Yonetani and Theorell (1964), described in Chap. 7.3.

Figure 36 illustrates these two procedures of evaluation, exemplified by additive = competitive inhibition of ADP plus ADP-ribose (Fig. 36a and c) and overadditive interaction between ADP and o-phenanthroline (Fig. 36b and d). DRCs of the observed inhibitory effects are shown in comparison with the graphical plots of inhibitor concentrations vs. the reciprocal of the reaction velocity of the enzyme alcohol dehydrogenase. Tabulated data were taken from Chou and Talalay (1981) to construct DRCs and the Yonetani–Theorell plots. Whereas the experimental points of the combined effects of ADP and ADP-ribose clearly match with the respective additive DRCs (Fig. 36a), the effects of ADP plus o-phenanthroline show that the experimental DRCs are not compatible with additivity (Fig. 36b).

In the Yonetani–Theorell plot, competitive interaction of mutually exclusive drugs is characterized by parallel lines, as obtained in Fig. 36c. Mutually nonexclusive drugs give nonparallel intersecting lines as in

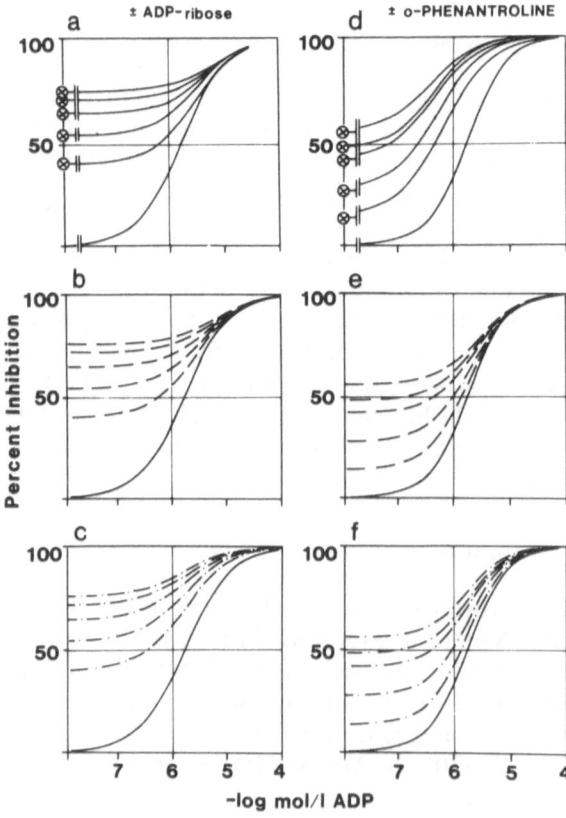

Fig. 37. Experimental (**a, d**) and theoretical DRCs (**b, e** and **c, f**) of inhibitory effects of ADP alone and in the presence of different concentrations of ADP-ribose (see Fig. 36c) and o-phenanthroline (see Fig. 36d). **a, d** DRCs correspond to the DRCs of Fig. 36a and b. Respective additive (**b, e**) and independent (**c, f**) DRCs. Experimental DRCs in **a** correspond to additive DRCs, whereas the experimental DRCs in **d** do not match with additive DRCs but exceed them. The DRCs in **d** are more similar to the independent curves in **f** – but even exceed those curves, also

Fig. 36d. It can be calculated that independent effects give a crossing at the horizontal line (not shown). Crossing above the horizontal line, as in Fig. 36d, indicates that o-phenanthroline has lowered the dissociation constant of ADP (Cleland 1970). This means that the combined effects must exceed independent effects in DRCs, as is observed in Fig. 36b by the left shift of DRCs, and in Fig. 37 by comparison of experimental DRCs (d) with curves of independent effects (f). The decrease in the ED_{50} of ADP by o-phenanthroline was dependent on the concentrations of o-phenanthroline, and amounted to a dose ratio = 2.7–4.

Where the combination differs from additivity we might be interested in the *type of interaction*. Figure 37d shows the DRC of the inhibitory effects of ADP in the absence and presence of o-phenanthroline, with the theoretical curves for additive (e) as well as independent combinations (f), displayed separately. For comparison, the analogous DRCs of ADP $\pm$ ADP-ribose with their theoretical curves are also shown (Fig. 37a–c). The combinations of ADP plus ADP-ribose (Fig. 37a) are less pronounced than the respective independent combinations (Fig. 37c), whereas the combined effects of ADP plus o-phenanthroline (Fig. 37d) exceed calculated independent effects (Fig. 37f).

When the experimental points are forced to be fitted to additive and independent DRCs, the following p-values in the F-test were obtained. The combined effects of ADP with ADP-ribose were not significantly different from additive (p = 0.11) but significantly different from independent (p = 0). Those of ADP with o-phenanthroline were significantly greater than additive as well as greater than independent (p = 0).

Kremer et al. (1980) have investigated the interaction of phenyl-phosphate and sulfate on the enzyme transketolase. They provided evidence for a competitive interaction at a single anion-binding domain. Reevaluation of their data gave DRCs of combined effects which appeared additive and gave parallel lines in the Yonetani–Theorell plot (not shown).

8.1.2 Physiology

Quite frequently, combined effects of endogenous substances in combination appear greater than independent combinations when acting at different sites, i.e., potentiated. At the same time they appear overadditive, due to relatively flat DRCs. Similar enhanced effects are also seen in time-course studies (e.g., Pöch 1983).

Figure 38 shows selected DRCs of combined effects of first and second messengers. Figure 38a shows that the increase in platelets' calcium concentration, induced by serotonin, was markedly enhanced by noradrenaline, which by itself produced a small effect. The increase in short-circuit current elicited by the cholinergic agonist, bethanechol, was markedly enhanced by the intracellular second messenger, cAMP, applied as 8-bromo-cAMP (Fig. 38b). This potentiation can be explained by the synergistic interaction of the two second messenger systems, cAMP and phosphoinositol-calcium (bethanechol) (Yajima et al. 1988).

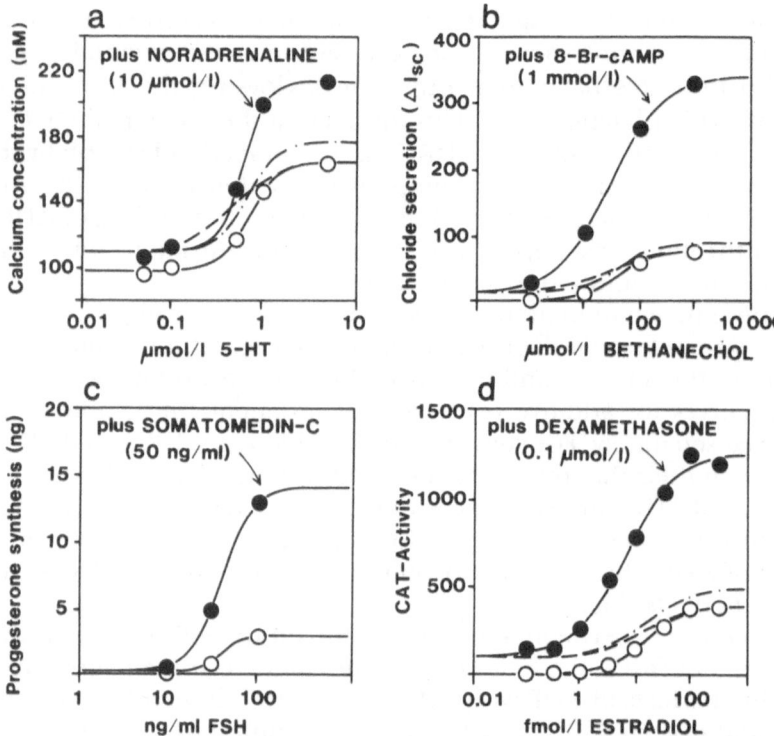

Fig. 38. Examples of combined effects of first and second messengers, respectively, in mediating physiological actions, expressed by DRCs, which were derived from the literature: **a** De Chaffoy de Courcelles et al. (1988), **b** Yajima et al. (1988), **c** Adashi et al. (1985), **d** Ankenbauer et al. (1988). The slopes of the DRCs were 2.7 (**a**), 1.4 (**b**), 3.8 (**c**), and 1.2 (**d**). The effects measured were: **a** increase in calcium concentration of blood platelets, **b** chloride secretion in guinea-pig colon (ΔI_{SC} in $\mu A/cm^2$), **c** progesterone accumulation in granulosa cells (ng/culture), **d** induction of chloramphenicol acetyltransferase (CAT) in MCF7 cells (pmol/min × mg). All effects in combination (●) exceed additive (broken line) as well as independent effects (broken dotted line) (not shown in **c**)

Threshold concentrations of somatomedin C obviously potentiated the accumulation of progesterone, elicited by follicle stimulating hormone (FSH) (Fig. 38c). Dexamethasone potentiated the increase in chloramphenicol acetyltransferase (CAT), elicited by estradiol (Fig. 38d). Hence, combination of an estrogen and a glucocorticoid resulted in a synergistic (potentiated) increase in transcription.

8.1.3 Pathophysiology

These examples show that in diverse areas of physiology, marked enhancement can be observed in combination. The same applies to pathophysiology. Only two examples are shown here (Fig. 39) with pronounced potentiation of immunological reactions. Figure 39a provides evidence for

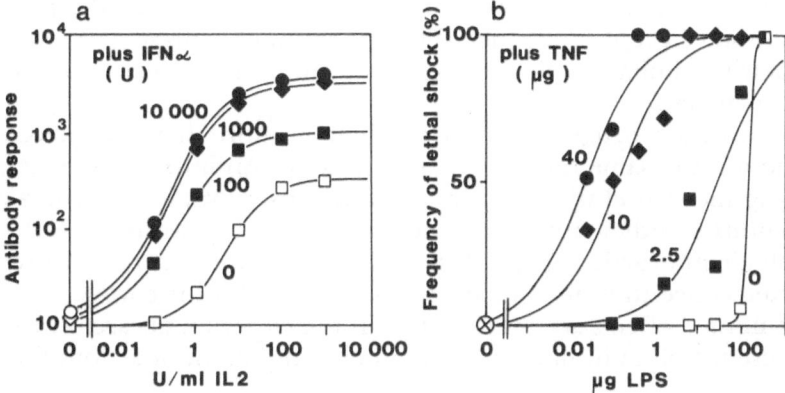

Fig. 39. Further examples of combined effects in physiology (**a**) and in pathophysiology (**b**), expressed by DRCs. Data of Delfraissy et al. (1988: fig. 1) on antibody response of IL 2 in the absence and presence of interferon (**a**), and of Rothstein and Schreiber (1988: table 1) on lethal shock produced by endotoxin lipopolysaccaride (LPS) ± tissue necrosis factor (TNF) (**b**)

a favorable potentiation of the IL-2-induced antibody response by interferon (IFN-α) (Delfraissy et al. 1988). The ED_{50} of IL 2 was decreased by IFN-α up to about 50-fold, and the maximum response to IL 2 was also considerably increased.

Figure 39b shows a deleterious potentiating response on lethal shock by tumor necrosis factor (TNF) on the action of lipopolysaccharide (LPS) (Rothstein and Schreiber 1988).

In summary, combined effects in biochemical and physiological or pathophysiological reactions appear of great importance, and numerous examples suggest pronounced enhancement of those reactions which are controlled by many factors, and, notably, different signaling mechanisms.

8.2 Experimental pharmacology

Many reversibly acting drugs either mimic or antagonize the effects of endogenous transmitters, hormones or other factors. Hence, we can expect – in certain cases – marked enhancement between drugs acting in the same direction, as observed in physiology. As a matter of fact, similar, if not the same, experiments can be described under the heading of biochemistry and physiology or pharmacology, e.g., the experiments with ADP and o-phenanthroline in Figs. 36 and 37, or the experiments with the synthetic glucocorticoid, dexamethasone in Fig. 38d.

8.2.1 Dose-response studies: site of action

However, there are certain aspects which justify describing combined effects in experimental pharmacology separately. Pharmacologists are often interested in the site of action of drugs, e.g., whether a given drug binds to a certain site (receptor) and acts through binding to this site. The situation is somewhat similar to biochemistry, yet often more complicated or difficult to analyze.

For instance, drugs not infrequently show more than one mechanism of action in the same range of concentrations. An interesting example is the drug nicorandil with two different actions, nitrate- and cromakalim-like actions. If either "nitrocompounds" (such as nitroprusside-Na) or potassium-channel activators (such as cromakalim) are tested in the presence of a fixed concentration of nicorandil in isolated bovine coronary arteries, the combined effects are overadditive. If the experiments are repeated in the presence of an inhibitor of the other component of action, the com-

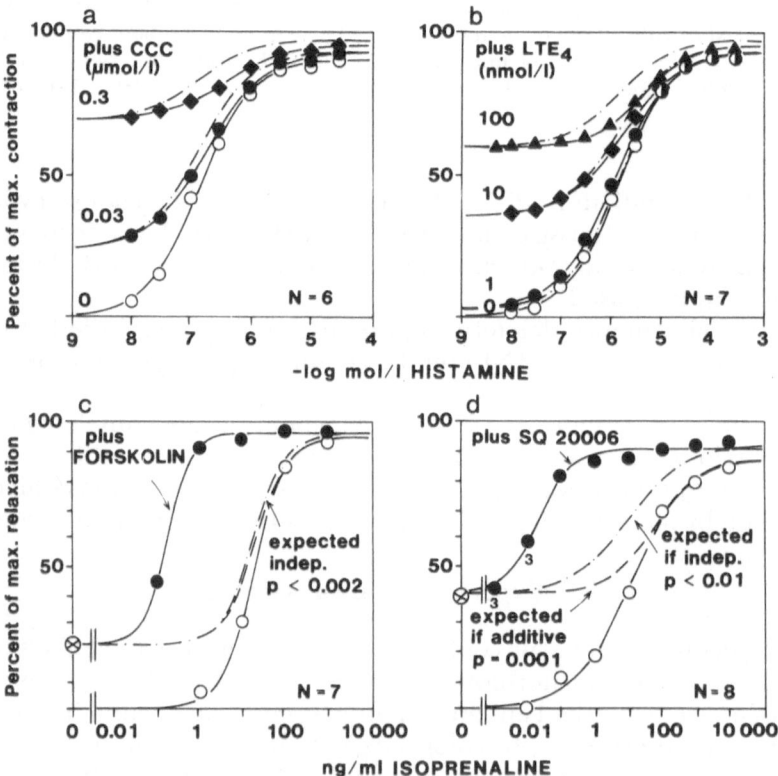

Fig. 40. Combinations of drugs with different actions. **a, b** Examples for two-receptor one-transducer mechanism. DRCs of histamine alone (○) and in the presence (filled symbols) of carbachol (**a**) and leucotriene E_4 (LTE$_4$) (**b**). Redrawn from results of Undem and Adams (1988). **c, d** Examples for "sequential" interactions. DRCs of the β-adrenergic agonist isoprenaline alone (○) and in the presence (filled symbols) of the adenylate-cyclase activator forskolin (40 nmol/l) (**c**), and the phosphodiesterase-inhibitor SQ 20006 (25 μmol/l) (**d**). Comparison with additive (broken line) and independent (broken dotted line) combinations by χ^2 goodness-of-fit test

bined effects appear additive (Kukovetz et al. 1991, Holzmann et al. 1992), just as with drugs which share one site of action, and apparently interact in a competitive manner (Fig. 22) (see Pöch and Holzmann 1980/1981, Pöch et al. 1990a).

The experiments mentioned above were done with more than one fixed concentration of nicorandil and showed that *significant* deviations from additivity occurred (under conditions in which nicorandil could exert both actions) if the differences were pronounced, i.e., with nicorandil concentrations which by themselves produced more than about 30% relaxation. Similar results were obtained with other combinations of drugs with different sites of actions. The explanation for this finding is that such combined effects often correspond to or slightly exceed effects of independent combinations. Hence, the differences between additive and independent interactions have to be marked in order to detect significant deviations of combined effects from additivity. Only where the combined effects also exceed the effects of independently interacting agents, can overadditive effects in pharmacology be proven more easily. Experimental examples are given in Fig. 40c and d.

Also, there are situations in which the theoretical response of additive interactions resembles that of independent actions in combination, as schematically indicated in Fig. 13c. This phenomenon occurs with drugs which possess steep DRCs, steeper than those of receptor agonists and antagonists. An experimental example is the interaction of two phospho-diesterase-inhibitors, theophylline and methylisobutylxanthine (MIX) (Pöch et al. 1990a). It is more commonly seen with cytostatics, the combined effects of which will be described in Chap. 8.4.

In these situations, only significant deviations from additivity as well as independence point to different sites of actions of the compounds tested in combination, whereas additive combinations matching independent interactions do not give any indication whether or not the agents share the same molecular site of action. This becomes clear when one considers that independent actions require different sites of actions.

8.2.2 Extent of enhancement

Another aspect of interest is the magnitude of combined effects of compounds acting at different sites. As pointed out in Chap. 2, independent actions can be expected for compounds acting at different sites *and* independently of each other. This simple situation is not expected to occur in general, since we know that different chains of reactions are often interrelated and activation or inactivation of one reaction may have a profound influence on others. The examples given above for physiological and pathophysiological reactions underline this point. Since many of these responses, to first and second messengers for example, lead to greater effects than expected for independently acting agents, we might expect greater than independent effects also to occur with drugs leading to a similar response by activation or inhibition of different reactions. However, this is speculation, and we should therefore investigate experimental results beforehand.

As an example, the combined inhibitory effects of all possible binary combinations of 6 drugs on platelet aggregation are shown in Table 5. These drugs all act by different mechanisms. The results have already been described and compared with effect-additive (Chap. 6.1.1) and independent combinations (Chap. 6.1.2). With the exception of 2 combinations, all others either showed combined effects equal to or greater than independent. Similar results have been obtained with many combinations of differently acting smooth muscle relaxants (not shown).

Nosál'ová et al. (1990) reported on single-dose studies with drugs to inhibit experimental gastric damage in rats, induced by phenylbutazone or stress. Ranitidine and pentacaine exerted greater effects than expected when acting independently, as evident from an inspection of the data. For instance, smaller doses of these drugs produced 54% inhibition of mucosal lesions, whereas 38% would have been expected for independent actions; higher doses inhibited lesions by 91%, compared with 83% calculated for independent combinations. Also, histamine-induced gastric secretion was inhibited to a similar extent in combination.

Hence, it appears that, in general, drugs with different sites of action show slightly greater effects than expected for independently interacting agents. However, in some cases weaker or much stronger effects in combination are seen.

Weaker effects in combination are found with drugs activating a "two-receptor-one-transducer system" (Undem and Adams 1988). In other words, they activate the same chain of reaction by binding to different receptors. The experimental results shown in Fig. 40a and b were taken from Undem and Adams (1988). These authors used the equation of Ariëns et al. (1956b), corresponding to Eq. (1), for calculation of independent effects but designated this type of interaction as "functional additivity".

Much greater than independent effects are observed with drugs interacting in sequence in the same chain of reaction, e.g., in the pathway of synthesis of tetrahydrofolate, termed sequential blockage (Black 1963). Experimental DRCs of this interaction with sufanilamide plus pyrimethamine are shown in Chap. 8.4. Similar interactions have been observed in the receptor–adenylate cyclase–cAMP–phosphodiesterase-(PDE)-pathway. It has been observed that the effects of receptor agonists leading to activation of adenylate cyclase are greatly enhanced by a direct activator of adenylate cyclase, forskolin (Seamon and Daly 1981). Figure 40c provides an example for this type of interaction, where the combined effects clearly and markedly exceed those of independent agents. Another interesting example is the interaction of activators of adenylate cyclase (isoprenaline) and PDE-inhibitors, as shown with papaverine earlier (Pöch and Holzmann 1980/1981, Pöch et al. 1990a). Figure 40d shows a similar marked enhancement of relaxant effects of isoprenaline in the presence of another PDE-inhibitor, SQ 20006 (see Kukovetz et al. 1976, Fredholm et al. 1979).

8.2.3 "Unspecifically" acting agents in combination

The aforegoing section described combined effects of specifically acting drugs, the effects of which can be attributed to binding to special sites at proteins, e.g., transmitter receptors. One facet associated with unspecific actions of lipophilic agents has already been touched in Chap. 2.1.4 concerning the possible mechanisms underlying the phenomenon of additivity. Most earlier studies have found combined effects of unspecifically acting anesthetics to correspond to or resemble additive combinations (e.g., Bürgi 1938, Rang 1960, DiFazio et al. 1972). Most of them employed isobolographic analysis of combined effects. This problem has been recently reinvestigated with lipid solvents in a model system with tracheal smooth muscles in vitro (Schwarzl 1992). Figure 41 illustrates the important findings obtained in these experiments. Some showed significant deviations from additivity with the solvents in combination (Fig. 41 b and d), some did not (Fig. 41 a and c). Similar results were obtained with the solvent DMSO in combination with a specific drug, salbutamol (Fig. 41 e and f). From these results we conclude (1) that the "unspecific" model substances do not interact competitively at one single binding site, (2) that the combined effects exceed the effects of independent agents and, therefore, yield a combined effect which phenomenologically often reflects additivity.

8.3 Clinical pharmacology – drug combinations

8.3.1 Dose-response studies

The evaluation of combined effects in healthy volunteers and patients is rarely based on complete dose-response studies. However, in some situations this is possible, and should, in the opinion of the author, be evaluated by DRCs more often.

For instance, frequency of occurrence of anesthesia has been evaluated by DRCs recently (Ben-Shlomo et al. 1990; Tverskoy et al. 1988, 1989). From these data, DRCs of frequency of observed anesthesia together with the curves of the theoretical combinations have been reevaluated, shown in Fig. 42a and b. As evident from the upper panel of Fig. 42, threshold doses of midazolam caused anesthesia in the presence of the opiate, fentanyl in 59%, and 21% in the presence of the barbiturate, methohexitone. When evaluated as described in Chap. 4, the observed effects in combination were significantly greater than independent in Fig. 42a and b but significantly differed from additive in Fig. 42a only, with $p = 0.003$ (chi-square goodness-of-fit test).

Also, dose-response studies with ethanol and metaclazepam on task performance (Dietz et al. 1984) were expressed by DRCs of observed effects together with theoretical DRCs in Fig. 42c and d. It appears that the deteriorating effects of ethanol were enhanced by metaclazepam roughly to the extent of independent acting agents (Fig. 42c) and vice versa (Fig. 42d). In order to obtain meaningful DRCs it was necessary to com-

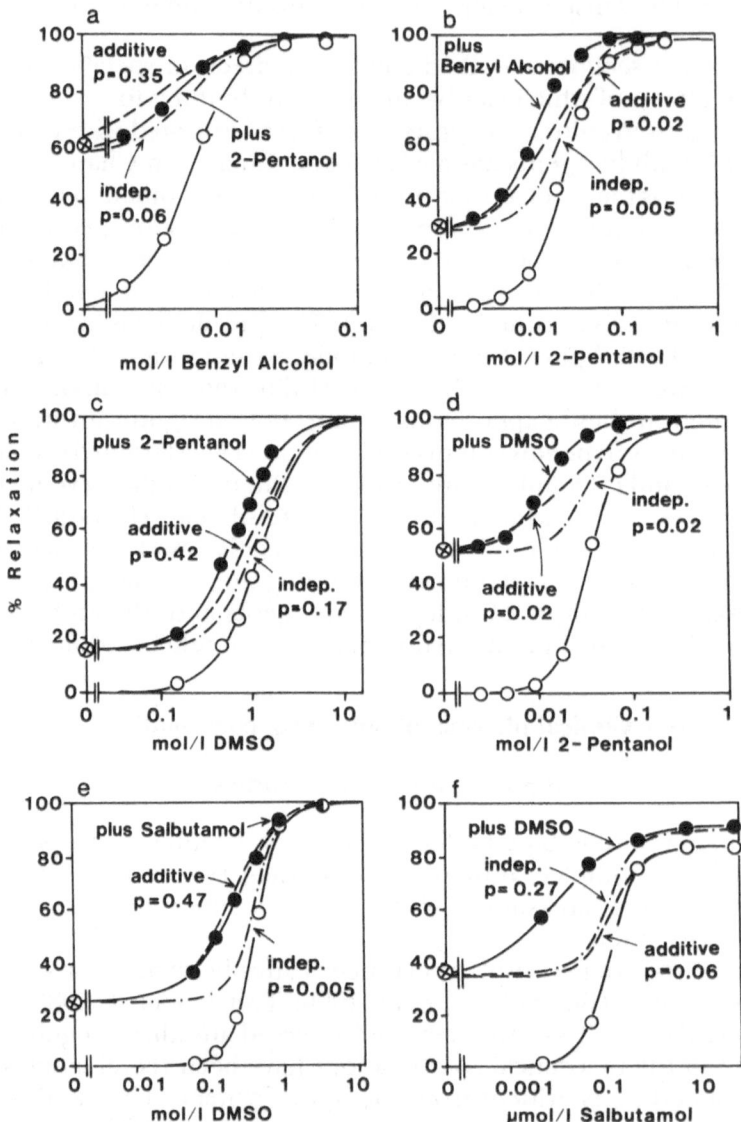

Fig. 41. Dose-response curves of relaxant effects of "unspecifically" acting lipid solvents and "specific" drugs. Results from N = 8–12 (**a–d**) and 12 (**e, f**) bovine tracheal muscle strips, precontracted by 27 mmol/l KCl: solvents in the presence of **a** 0.01 mol/l 2-pentanol, **b** 0.005 mol/l benzyl alcohol, **c** 0.009 mol/l 2-pentanol, **d** 0.42 mol/l DMSO, **e** 0.04 μmol/l salbutamol; **f** salbutamol in the presence of 0.35 mol/l DMSO. The slope values of the DRCs of the agents when tested alone in **a–f** were: 2.3, 2.0, 1.9, 2.5, 2.6, and 1.5. Comparison of observed effects in combination with theoretical DRCs. p-Values obtained by χ^2 goodness-of-fit test are indicated

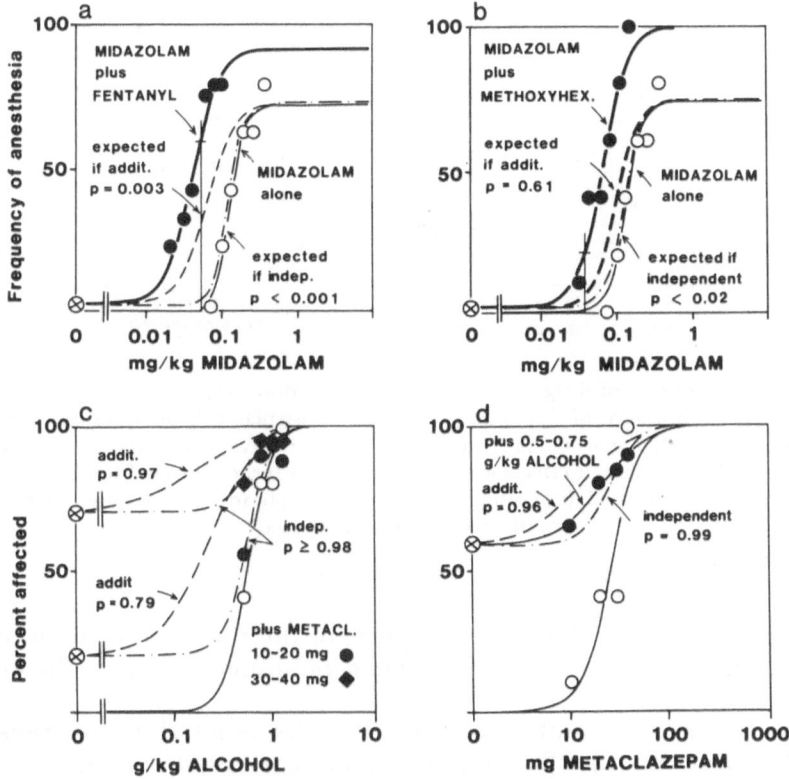

Fig. 42. Dose-response studies of frequency of anesthesia (**a, b**) and of deteriorating effects in humans of alcohol and drugs (**c, d**). **a, b** DRCs were constructed by ALLFIT from DRCs of Ben-Shlomo et al. (1990) (**a**) and Tverskoy et al. (1989) (**b**). Midazolam was given to 10 women alone (○) or together (●) with 1.9 µg/kg fentanyl (**a**) or 0.225 mg/kg methohexital (**b**). **c, d** DRCs were constructed from data of Dietz et al. (1984: table 1) from impairment of performance in the two test systems (N = 7–10). Comparison of observed with calculated DRCs for additive (broken line) and independent combinations (broken dotted line), including χ^2 goodness-of-fit statistics. The agents when tested alone exhibited slope values in **a–d** of 3.9, 3.9, 3.6, and 3.0

bine the results obtained with 10 and 20 mg as well as 30 and 40 mg metaclazepam, respectively (Fig. 42c). Similar grouping had to be done with the doses of ethanol to obtain the results shown in Fig. 42d. The combined effects were not statistically different from independent as well as from additive combinations. However, the latter should be considered as phenomenon, not as expression of a competitive interaction, analogous to additivity of organic solvents, discussed above.

8.3.2 Drug mixtures: dose-response studies

Therapy with drug combinations is often done by using mixtures of drugs with fixed-dose ratio combinations. Again, dose-response studies are rarely

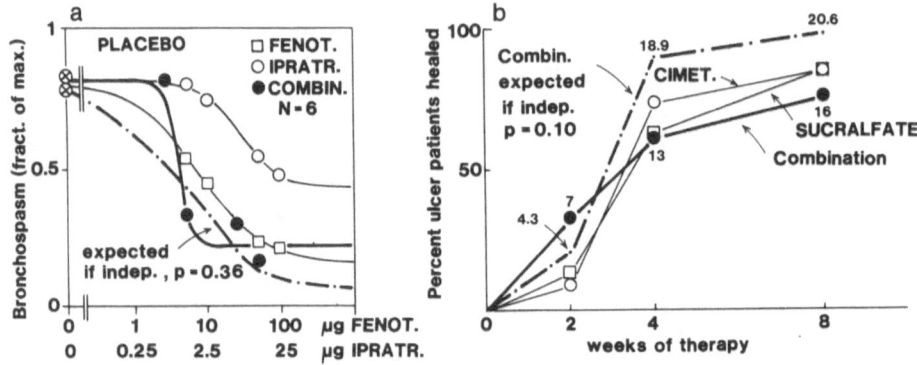

Fig. 43. Clinical effects of drug mixtures, expressed as dose response and time course. **a** DRCs of the inhibition of bronchospam were constructed from data of Schultze-Werninghaus (1981), **b** time course of ulcer healing response rate from data of Van Deventer et al. (1985). The combined effects were compared with those expected for independent interactions (broken dotted line), using the χ^2 goodness-of-fit test, and did not differ significantly from them, as indicated

performed. General, biometrical, aspects of the therapeutic application and analysis of drug mixtures have been published recently (Roebruck et al. 1990).

One empirical example is shown in Fig. 43a, which shows the relief of bronchospasm by the β-adrenergic stimulant fenoterol and by the muscarinic antagonist ipratropium, alone and in a 4:1 combination. The effect of the mixture was not significantly different from an expected independent response (Fig. 43a).

8.3.3 Drug mixtures: time-course studies

Time-course studies are more frequently done than dose-response trials in clinical studies. One example in the literature (Van Deventer et al. 1985) is evaluated here showing the time course of the frequency of ulcer healing in patients with the H_2-blocker cimetidine and with sucralfate, alone and in a fixed, almost equieffective combination (Fig. 43b). Again, the response to this drug combination was not significantly different from an expected independent response.

8.3.4 Single-dose studies

Most of the clinical studies with drug combinations were so far performed with single doses of two or more drugs in combination. These studies in part investigated the effect of the single drugs alone, which allows a comparison of observed combined effects with expected effects of independently acting drugs in frequency studies and in studies where independent effects can also be calculated (see Chap. 2.3.1). This approach is directed to the *magnitude of response* to drug combinations; other conclusions should not be drawn or with great reservations, due to the lack of DRCs. Nevertheless, such trials give valuable informations for therapy with single drugs and with drug combinations. Reports from the literature suggest that treatment with mixtures of drugs often results in slightly, or sometimes markedly, greater effects than expected for independently acting drugs.

Although a vast number of clinical reports describe effects of drugs in combination, many of them cannot easily be compared with results of other single-dose studies (see Chap. 6). A few clinical examples, however, have been described and discussed in Chap. 6. Tables 7 and 8 list additional examples of clinical studies with healthy volunteers and patients, which have been split into those describing effects (Table 7) and changes in control values (Table 8). Most of the studies show slightly greater effects in combination than those calculated for independently interacting drugs, or effects which nicely match with the calculated effects. In the clinical study of Kim et al. (1987) (Table 8), the reduction of frequency of arrhythmias to 10% was achieved with less of each drug in many patients, and none received more than in single dose therapy. Hence, a greater reduction would have been expected with full doses in combination.

There is one report in the literature which appears not to fit into this picture. According to Kayanakis and Baulac (1987), the combination of captopril and hydrochlorothiazide normalized blood pressure in 78%. From the individual rate of responders (72.5 and 67.5%, respectively), 91% was

Table 7. Combined effects (E_{A+B} *obs.*) in clinical studies with drugs A and B. Percent of maximum response compared with independent combinations (*indep.*) calculated from individual effects, E_A and E_B

A	B	E_A	E_B	E_{A+B} obs.	indep.	Response and clinical study
Timolol	Hydrochloro-thiazide	40	65	86	79	DBP < 90 mm Hg (12 weeks, 61 pat.) Leon and Hunnighake (1983)
Bendroflu-methiazide	Nadolol	46	49	85 (W)	72	DBP < 90 mg Hg (12 weeks, 140 white
		46	31	84 (B)	63	[W], 178 black subj. [B]) Freis et al. (1983)
Captopril	Hydrochloro-thiazide	46	54	88 (W)	75	DBP < 90 mm Hg (6 weeks,
		31	53	67 (B)	68	100 white [W], 98 black pat. [B]) Weinberger (1985)
Atenolol	Nitrendipine	25	20	70	40	DBP ≤ 90 mm Hg (4 weeks, 60 pat.) De Divitiis et al. (1984)
Pirenzepine	Cimetidine	58	69	89 (V)	83	inhibition of acid secretion (8 volun-
		34	63	90 (P)	76	teers [V], 5 ulcer patients [P]) Londong et al. (1980)
Pirenzepine	Ranitidine	42	80	97	88	inhibition of acid secretion (8 volunteers) Londong et al. (1981)

Table 8. Combined effects in clinical studies, analogous to Table 7. Percent of control values
(*PC*)

A	B	PC_A	PC_B	PC_{A+B} obs.	indep.	"Response" and clinical study
Low-dose Heparin	Dihydro- ergotamine	51 24	48 24	27 (S) 7 (G)	25 6	postoperative thromboembo- lism in 1480 surgery (S) and 454 gynecology (G) patients Buttermann et al. (1977)
Pirenzepine	Cimetidine	42	26	9	11	recurrent bleeding from gastro- duodenal lesions (34 patients) Londong et al. (1982)
Nitroglycerol	Pindolol	60	28	16	17	ischaemia-induced ST- depression (19 angina patients) Kaltenbach et al. (1970)
Quinidine	Tocainide	33	39	10	13	frequency of ventricular pre- mature complexes (20 patients) Kim et al. (1987)
Terfenadine (30 mg b.d.)	Ranitidine (600 mg b.d.)	54 42	89 86	42 (W) 18 (E)	48 36	histamine-induced wheals (W) and erythema (E) of skin Paul et al. (1988)

calculated for independent combinations. The study involved 211 patients
with mild to moderate hypertension, and there was a rather high value of
45% of patients which showed normalization of blood pressure with place-
bos. Even the placebo-corrected responder rate (see Chap. 2.3.1) of 60% in
combination is below the calculated rate of 70.5% for an independent action. However, there were only 0.5% *non*-responders of those receiving the
combination compared with 7% calculated for independent interaction
(27.5 and 25% were non-responders with single-drug therapy). In this respect, the combination was more effective than independently acting drugs.

From these results and from theoretical considerations it is clear that
we can expect, in general, greater therapeutic efficacy with drugs combined with respect to independent combinations. The differences of the
response to mixtures compared to single drugs is expected to be greatest
when the drugs are applied in about equieffective and about half-maximal
effective doses, as can be calculated for independent effects. This is easily
demonstrated by two examples. If A and B each produce 50% response, the
independent response in combination is 75% ; if A produces 50% and B
elicits 10% response, the combined effect of an independent combination
is only 55%.

8.4 Chemotherapy of infections and tumors

For many scientists and clinicians this section may be the most interesting
one. Experimental studies and therapeutic applications of drug combina-

tions have been and still are very important. Hence, many reports concerning combined effects come from this field, and some books have been published which exclusively deal with drug combinations in chemotherapy (e.g., Klastersky and Staquet 1982, Chou and Rideout 1991). This topic cannot be completely covered in this book alone. However, the new approach can be demonstrated by a few examples.

One of the most interesting aspect of chemotherapeutic agents is that their DRCs are generally steep. In consequence, the theoretical DRCs for additive and independent combinations do not differ greatly, and often coincide. This situation is schematically illustrated in Figs. 8 and 13c and by an experimental example in Chap. 8.4.2. Also, quite often, combined *effects* of chemotherapeutic agents (Tsai et al. 1989, Chou and Talalay 1984) are not much different from additive or independent combinations, the latter illustrated for cisplatin and etoposide in Figs. 34c and 35c.

8.4.1 Therapy of infections

There are many possibilities of enhanced efficacy of chemotherapeutic agents applied in combination (e.g., Berenbaum et al. 1982, Klastersky and Staquet 1982). It is not possible to allow a detailed presentation and discussion in this chapter, which emphasizes the application of the new approach. As outlined above, independent combinations are often as effective as additive interactions. Possibilities to compare the results of combined treatment with independent actions have been given schematically for dose-response studies in Fig. 8, and for time-course studies in Fig. 32a, including an experimental result with sulfamethoxazole and trimethoprim (Fig. 32b).

DRCs, as suggested in this book, are presented for the antimalarial effects of sequential blockers of tetrahydrofolate synthesis (Fig. 44a–c), and for chlorguanide plus pyrimethamine (Fig. 44d–f), based on the data of Rollo (1955). It is clearly evident that the antiparasitic effect of sulfanilamide is greatly enhanced by pyrimethamine (PYR), even by threshold doses (Fig. 44a). A comparison with additive and independent DRCs shows that the observed combined effects with this combination greatly exceed both, additive and independent combinations (Fig. 44a–c). On the other hand, the effect of chlorguanide is much less enhanced, similar to additive as well as independent combinations expected (Fig. 44d–f). This finding is in accordance with the isobologram analysis performed by Rollo (1955), showing roughly an additive response to this combination.

Prichard and Shipman (1990) described the antiviral effects of 2-acetylpyridinesemicarbazone (APTS) in combination with various concentrations of acyclovir (ACV). The DRCs have been recalculated by ALLFIT and are shown in Figs. 34a and 35a. The corresponding combined-effect graph in Fig. 34b shows that – at the 10% effect-level – the effects of APTS were somewhat more enhanced than expected for independently acting agents. On the basis of the individual points, the observed combined effects were scattered around the line expected for agents acting independently (Fig. 35b). The overall combination, therefore, corresponds to in-

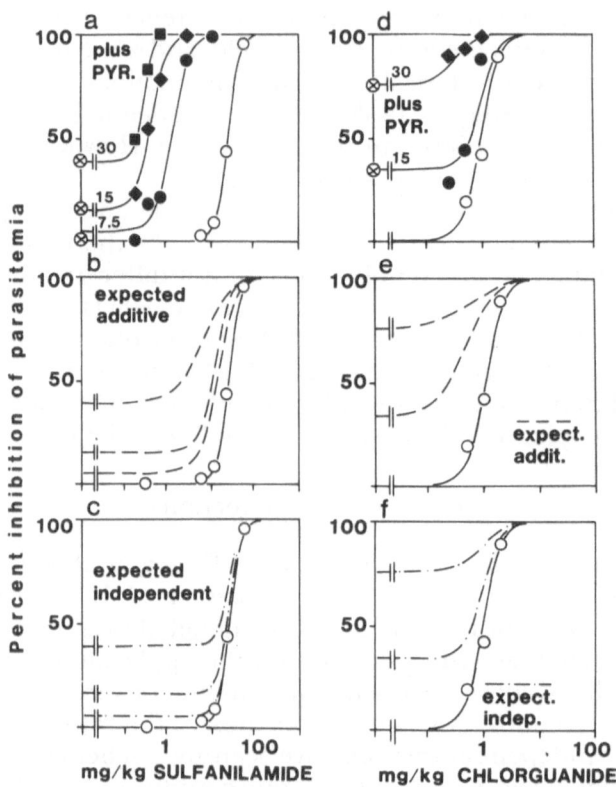

Fig. 44. Examples of antiparasitic agents in combination (Rollo 1955), expressed by DRCs of the inhibition of parasitemia induced by sulfanilamide (**a–c**) or chlorguanide (**d–f**) in the absence (○) and presence (filled symbols) of 7.5 to 30 μg/kg pyrimethamine, observed (**a, d**) and expected for additive (**b, e**) and independent combinations (**c, f**). Whereas the combined effects of sulfanilamide were markedly potentiated by pyrimethamine (**a**), the effects of chlorguanide in combination (**d**) resemble independent or slightly greater than independent effects

dependent effects between APTS and ACV or slightly greater effects than independent.

Recent results from Pancheva (1991) show that the antiviral effect of ACV is potentiated by ribavirin by an interesting mechanism of sequential blockage, i.e., by removing an antagonist of ACV. Due to a reduction of dGTP by ribavirin, an endogenous competitor of ACV-TP formed from ACV, the constraint for the action of ACV is reduced. Although not analyzed by Pancheva (1991), it is obvious that the observed effects in combination greatly exceed effects of independent combinations, hence represent true potentiation, on the basis of the following considerations. The isobolographic analysis shows isoboles of overadditive interactions whith the greatest distance from the additivity line at about 1/4 of the ED_{50} of ACV and ribavirin, applied separately. Since the slopes of the DRCs of ACV are about 3.5 and 2.8, and for ribavirin about 3 and 1.7 for the two virus strains investigated, the ED_{30}s correspond to about 2/3 to 3/4 of the ED_{50}s. Hence, the zenith of isoboles of *independent* actions clearly correspond to underadditive combinations (as determined by Pöch et al. 1990c). Therefore, overadditive isoboles represent combined effects greater than independent, i.e., potentiation.

8.4.2 Tumortherapy

There is a vast amount of reports of combined effects of antitumor drugs in the literature (see Chou and Rideout 1991). Typically, antitumor drugs possess DRCs with slopes between 1 and 2. This implies that additive combinations often coincide with independent combinations, i.e., with drugs exhibiting steep DRCs with slopes around 1.6 (see Figs. 13c and 45), (Unkelbach and Pöch 1988, Pöch et al. 1990a). This situation is also illustrated by the example of cisplatin and BM 41440 in Chap. 10.

As with many other chemical agents in combination, described above and below, the combined effects of chemotherapeutic drugs often slightly exceed or correspond to independent interactions (e.g., Fig. 35d). Therefore, they often correspond to "effect-multiplication" in linear-dose log-response lines (Fig. 8) (Berenbaum 1981).

Some combinations show clear enhancement above independent effects, e.g., P-30 plus tamoxifen (Mikulski et al. 1990). Figure 45 shows this pronounced enhancement expressed by DRCs of cytotoxic effects of the P-30 protein and the anti-estrogen, tamoxifen, on human pancreatic carcinoma cells.

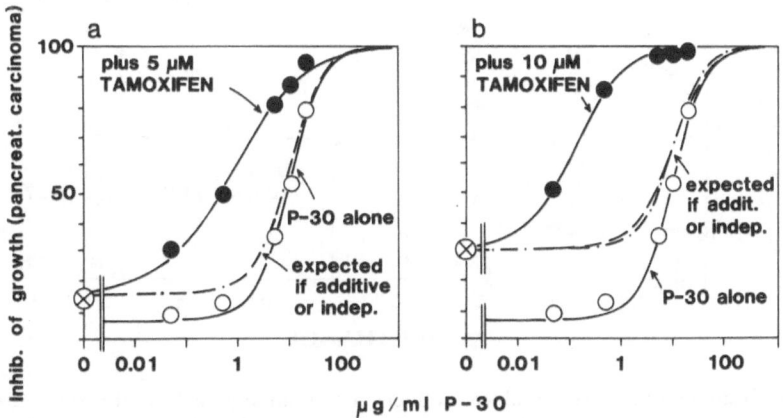

Fig. 45. Examples of experimental combined effects of antitumor agents. Percent inhibition of growth of pancreatic carcinoma cells by P-30 protein alone (○) and in the presence of 5 (**a**) and 10 μmol/l tamoxifen (**b**) (●). DRCs were constructed from data of Mikulski et al. (1990: table 3). The DRC of P-30 alone had a slope of 1.36. The calculated curves for additive and independent combinations match, and the experimental effects significantly differ in the χ^2 goodness-of-fit test from both theoretical curves

8.5 Experimental and environmental toxicology

Among the fields where effects in combination are of interest and importance, toxicology is probably the leading discipline. Many "classical" papers
are related to toxic actions (Bliss 1939, Finney 1942, Plackett and Hewlett
1952, Ashford and Smith 1964), and some books deal with toxic effects in
combination (e.g., Hewlett and Plackett 1979, Brunden et al. 1988,
Goldstein et al. 1990). Also, a recently published book on mixtures of
chemicals mainly deals with toxic actions (Vouk et al. 1987).

Besides toxicodynamic mechanisms, we have to consider interactions
which lead to changes in the concentration of drugs or xenobiotics, e.g., by
enzyme induction or inhibition (King and Parke 1987). We do not have

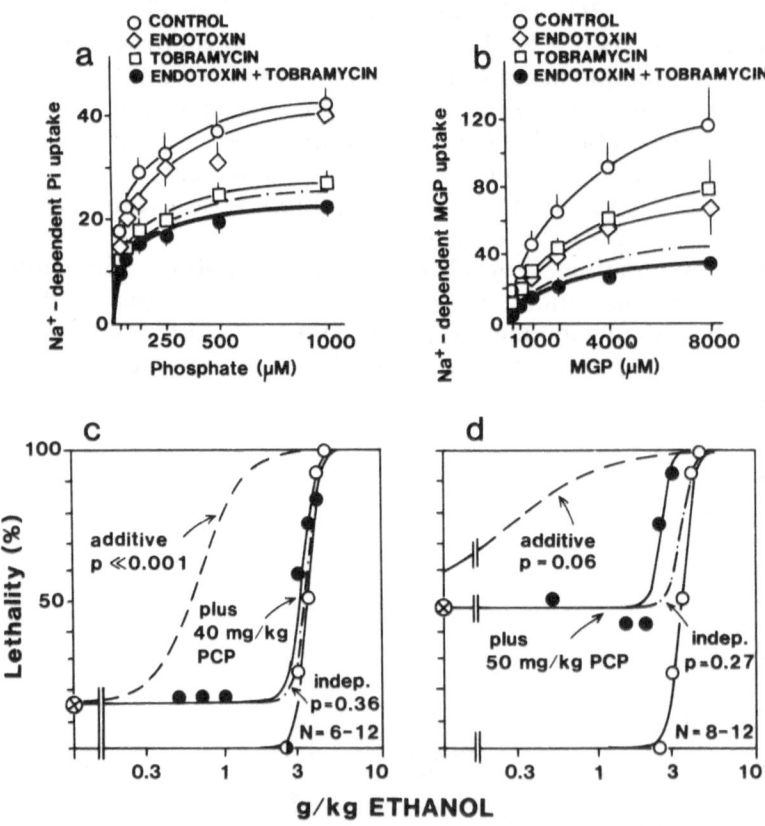

Fig. 46. Examples of experimental toxic effects in combination. **a, b** Nephrotoxic effects of
tobramycin and endotoxin on sodium-dependent uptake of phosphate (**a**) or α-methylglucopyranoside (MGP) (**b**), alone and combined. Redrawn from Joly et al. (1991: figs. 1 and 2).
c, d Lethality of ethanol and phencyclidine (PCP). DRCs of ethanol alone (○) and in the
presence (●) of 40 mg/kg (**c**) and 50 mg/kg PCP (**d**). Ethanol alone exhibited a slope = 11.2–
11.6. Comparison of observed effects in combination with DRCs for additive (broken line)
and independent (broken dotted line) combinations (χ^2 goodness-of-fit test). Redrawn from
Pöch et al. (1990b)

general models to quantitate changes in effects associated with toxico-kinetic interactions, therefore, we have to concentrate on toxicodynamic interactions (of independent actions).

Typically, we observe very steep DRCs with slope-values between 2 and 10 when studying toxic actions. As a consequence, combined effects often show the phenomenon of additivity or underadditivity when they correspond to or exceed (!) independent effects. A comparison of combined effects with additivity appears appropriate where we are interested in the doses for a given effect or where we have reason to believe that toxic effects of substances might be due to binding to a common site.

Often, toxic effects in combination are studied only with one or very few doses of the agents. So, they can only be evaluated as single-dose studies, usually with respect to independence in action, as described in Chap. 6.1.2. These studies show that, in general, combined effects exceed, or correspond to, independent effects, as shown recently for neuro-depressant and lethal interactions of ethanol and other drugs (Figs. 15 and 46c and d) (Pöch et al. 1990b, c). In the same experiments, the relationship to additivity varies greatly, often, combined effects appear underadditive, as in Figs. 15 and 46c and d.

In order to protect workers and consumers from toxic mixtures of chemicals as well as the environment from water pollution, threshold limits, food and water residue levels, and water quality criteria were enacted or proposed mainly on the basis of additive combinations (see Altenburger et al. 1992). Although we cannot expect, as a rule, that toxic agents truly act like dilutions of other components in mixtures, it appears justified to rely on the additivity model for calculating the doses in combination from a pragmatic point of view.

By this means *greater* than independent effects in combinations, which frequently occur, can be taken into account. Such combined effects may phenomenologically often correspond to underadditive but also to additive combinations, referred to above. Hence, it appears appropriate to consider additivity for calculation of threshold doses in mixtures.

The following example pictures the differences in the doses of additive and of independent combinations. The dose-response relationship is given for chemicals applied singly as well as in mixture with 2 and 4 equieffective components (n = 2 and 4, respectively) on the basis that DRCs of the components each have slopes = 3. A threshold dose = 1, and a threshold effect = 1% has been chosen arbitrarily.

Table 9. Threshold effect in additive and independent combinations

Condition	n = 2		n = 4	
	dose	effect (%)	dose	effect (%)
Applied singly	1×0.5	0.125	1×0.25	0.016
Combined if additive	2×0.5	1	4×0.25	1
Combined if independent	2×0.5	0.25	4×0.25	0.06

Table 9 shows that for DRCs with slope = 3, half the threshold dose produces 0.125%, and one quarter of it only 0.016%. *Additive* combinations of 2 times half the threshold dose, and 4 times the quarter of it, again produce the threshold effect, according to the model. Since the agents applied at 0.5-fold and 0.25-fold threshold doses produce less than half and less than a quarter of the threshold effect (1%), respectively, the combined effects in *independent* combinations are much below 1%. Hence, greater than "additive doses" are expected to produce 1% effect.

8.5.1 Examples of toxic interactions

Nephrotoxicity

Let us first look at examples. Joly et al. (1991) described what they call an additive toxicity of endotoxin and tobramycin on renal tubular cells in vitro. The reason for this investigation was to test the combined effects of endotoxin (derived from gram-negative bacterial infections) and aminoglycoside-antibiotics to treat such infections. The latter are known to be nephrotoxic in higher doses.

One of the most interesting results was published as DRCs, redrawn in Fig. 46a and b. It shows the sodium-dependent uptake of phosphate and α-methylglucopyranoside (MGP), respectively, as a function of the concentration of phosphate (Fig. 46a) and MGP (Fig. 46b) – in the absence and presence of one or the two agents combined. The DRCs in presence of tobramycin plus endotoxin were slightly below the calculated DRCs of independent effects. Hence the inhibitory toxicodynamic effects in combination slightly exceeded the calculated effects of independent interactions. This result shows that toxic concentrations of both compounds contribute to the toxic effect, slightly more than on the basis of the toxic effects of each of them, conforming to independence.

Hepatotoxicity

Many chemicals are hepatotoxic, including ethanol and haloalkanes. A look into the literature reporting on combined effects reveals that the toxic actions of well known hepatotoxic agents are drastically enhanced by others, termed potentiation (e.g., Plaa et al. 1982, Pessayre et al. 1982, Mehendale 1984) or synergism (Bosma et al. 1988). Since in some of these experiments one compound enhances the effect of others even in minimum or non-effective doses, their action is also described as sensitizing (e.g., Bosma et al. 1988, Mehendale 1984). Various mechanisms are discussed, including perturbations in intracellular calcium levels (Agarwal and Mehendale 1984) and an enhanced bioactivation, i.e., oxidation of carbon tetrachloride (Mehendale 1984). Hence, besides toxicodynamic effects in combination, toxicokinetic effects play a role.

Changes in the enzymatic metabolism of xenobiotics can also explain protective effects of other agents in combination. Chadwick et al. (1984) have reported on a protective effect of lindane against chlorobenzene-induced hepatotoxicity and have explained this effect by enhanced inactivation of the toxic metabolite, chlorobenzene-3,4-epoxide after pretreatment of lindane.

Another example of hepatotoxicity is the interaction of cadmium and chloroform, investigated by Stacey (1987). Inspection of his data shows greater than independent effects on intracellular potassium concentration. For instance, 25 µmol/l cadmium reduced potassium by 8%, while 15, 30, and 60 mmol/l chloroform decreased the potassium concentration by 9, 17, and 67% applied alone. In the presence of 25 µmol/l cadmium, chloroform caused a reduction by 49, 62, and 79%. The calculated independent effects are 17, 24, and 70%.

Lethality

As already mentioned, lethal interactions appear (in general) to correspond to greater than independent interactions, usually slightly greater. Examples are given in Fig. 46c and d. It is interesting that the combined effects appeared underadditive.

On the basis of the steepness of the DRCs for the lethal effect of different substances and the relationship between additivity and independence (see Fig. 13), we can expect slightly greater than independent effects to correspond to either overadditive, additive or underadditive combinations. Therefore, it is not too surprising to learn that a large proportion of lethal interactions of 27 industrial chemicals, in all possible pairs, fit the model of additivity (Smyth et al. 1969).

In some cases the combined lethal effects are much greater than expected for independent interaction. These examples include pentazocine plus tripelennamine (Waller et al. 1980), methylxanthines plus isoniazide (Harris et al. 1986), and propranolol plus morphine (Winter 1974).

Carcinogenesis

One of the most important questions in environmental toxicology is how carcinogens interact (Reif 1985, Künstler and Klein 1989, Schmähl 1988), how they jointly raise tumor incidence and shorten the latency period, respectively (Reif and Colton 1984). There is no doubt that we are all confronted with more than one carcinogen.

One of the most careful studies was reported by Elashoff et al. (1987) and by Berger et al (1990). Some of the combinations reported by Elashoff et al. (1987) could be expressed by DRCs of one cancerogen in the presence of a fixed dose of another one. Four such DRCs are shown in Fig. 47. Whereas all combinations exhibited a tumor incidence at least as great as the one expected for independently interacting agents, the relationship between observed and additive incidence is less clear. It appears from Fig. 47b and c that the combinations behaved as an additive interaction, including the response to CYCA in the presence of LAS. The latter pair in reversed order, i.e., LAS in the presence of CYCA, markedly deviates from an additive interaction (Fig. 47d). Hence, it is likely that the phenomenon of additivity with CYCA plus LAS, and probably also with DPN plus CYCA (Fig. 47c), occurred by mere incidence with additivity due to slightly greater than independent effects. Elashoff et al. (1987) compared their results with the model of independence and concluded that all but one combination fit that model. The mentioned exception is the combination of CYCA and LAS.

The tabulated frequency of liver tumor used allowed calculation of the corresponding independent effects, and thus comparison of observed and independent tumor rates (Fig. 48b). As can be seen from this figure, the observed incidence of tumors is mainly greater than expected for independently interacting agents, except at those above 90% expected. Further, more than half of the combinations showed a response which was greater than the sum of the individual effects.

In Chap. 6 we have discussed the results of many investigations listed in a data base on binary effects of chemical carcinogens (Arcos et al. 1988). This list shows a similar incidence of tumors as in Fig. 48b, i.e., greater incidence than the sum of the individual effects in 403 cases, 16 corresponding to the sum, whereas 249 cases showed a less than effect-additive

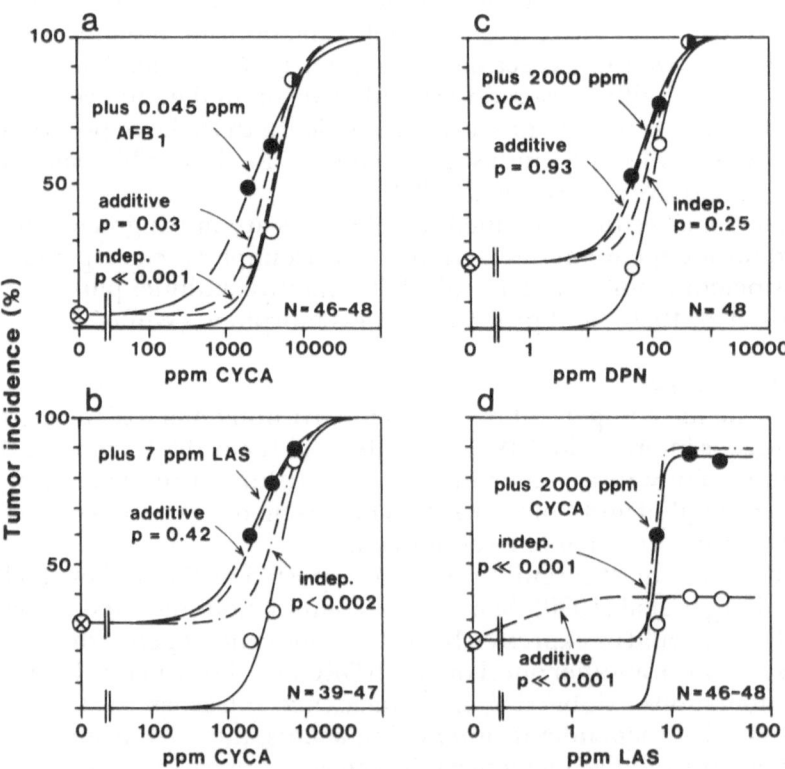

Fig. 47. Examples of environmental toxic effects in combination: incidence of liver tumors in male and female Fischer 344 rats. DRCs were derived from tabulated data of Elashoff et al. (1987) for the carcinogens alone (○) and in the presence of a second carcinogen (●). *AFB1* Aflatoxin B1, *CYCA* cycad flour, *DPN* dipentylnitrosamine, *LAS* lasiocarpine. Slope values for the agents when tested alone were 2.3, 1.9, 2.3, 2.0 in **a–d**. Comparison of observed and calculated tumor incidence was done by χ^2 goodness-of-fit statistics

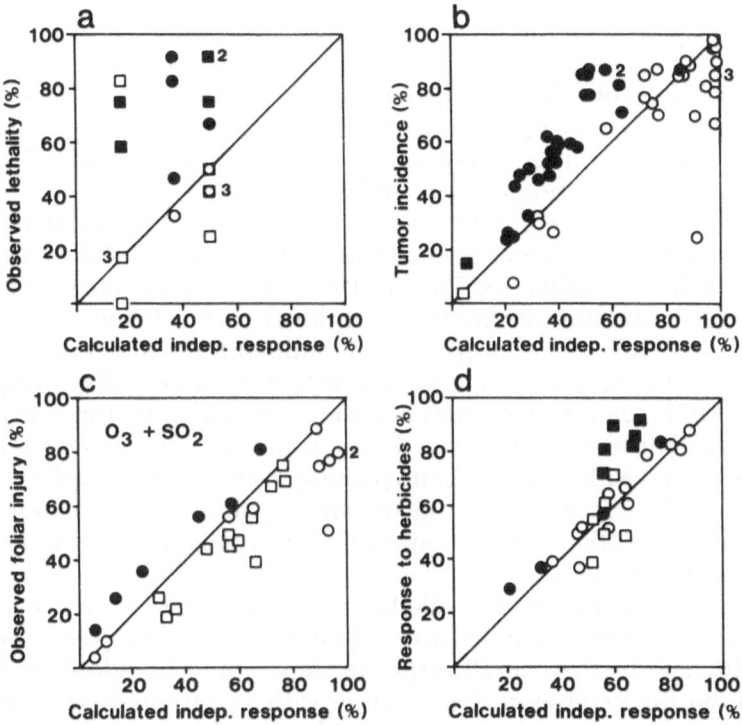

Fig. 48. Experimental effects in combination (y-axis) compared to calculated independent response (x-axis). ○ □ Combined effects equal to or smaller than the sum of individual effects, ● ■ greater than effect-additive. **a** Combined lethal effects of flurazepam and ethanol (circles), and of phencyclidine and ethanol (squares), derived from Pöch et al. (1990b); **b** Tumor incidence of combinations of various liver carcinogens, derived from tabulated data of Elashoff et al. (1987) (circles) and of Berger et al. (1990) (square); **c** Foliar injury of combined exposure to ozone and sulfur dioxide, derived from data of Heagle and Johnston (1979) (circles), and of Shertz et al. (1980) (squares); **d** Reduction in fresh weight of weeds by atrazine and alachlor (circles), and of atrazine and Sun 11E (squares), derived from data of Akobundu et al. (1975)

frequency. The numbers represent the total sum of the groups I to X, and some have been listed in more than one group. However, the principle result is that more than half of all studies (60%) gave greater effects than calculated by addition of the effects. As pointed out in foregoing chapters, a greater than effect-additive response indicates a greater than independent response, i.e., potentiation. A weaker response may or may not be explained by independent action in combination.

Teratogenesis
Quite recently, extensive studies with binary mixtures (Dawson and Wilke 1991a–c) as well as with ten agents (Dawson 1991) were performed with the frog embryo assay (FETAX). From these studies, evaluated on the basis of toxic units and isobolograms, respectively, it appears that those toxic

agents which are likely to act by the same molecular mechanism showed toxic units close to 1 (additive) and an additive interaction in isobolograms. For example, semicarbazide and β-aminopropionitrile inhibit lysyl oxidase and thereby inhibit the proper polymerization of developing collagen and elastin fibers. Others, with differing sites of action, in combination exhibited underadditive isoboles, termed "response addition", which might correspond to independent action.

It has been proposed that valproic acid produces microcephaly by interference with folate metabolism, a mechanism possibly shared by other aliphatic carboxylic acids (see Dawson 1991). Ten acids at doses of $1/10$ of the ED_{50} gave toxic units in combination close to 1, i.e., behaved additively. However, the DRCs of microcephaly-induction differed in steepness, as was determined by ALLFIT from the published dose-response frequency (Dawson 1991), with slope values between 2.49 and 7.46. Dawson (1991) also discussed differences between the acids and concluded that additional effects occurred with some of them, possibly accounting for differences in slope. It would not be surprising to obtain additive combinations with teratogens which share a common action but which are likely to bind (in part) to other sites as well.

Other actions, if present, would be expected to yield greater than independent effects which tend towards additivity with agents possessing very steep DRCs. Thus the overall response then would be additivity due to competition for a common site, together with a component of combined effect which phenomenologically resembles additivity.

8.6 Epidemiology – risks

In epidemiology, the risk of adverse events, and sometimes of protective factors, is investigated. For example, one is interested to see how certain events are influenced by risk increasing (or decreasing) factors. Several equations have been published for evaluation of combined effects of risk factors (e.g., Rothman 1974, Walter and Holford 1978, Stara et al. 1987). Many equations are given for models which assume independent interactions between risk factors and between risk factors and "spontaneous" or basal risk, observed in the absence of known risk factors. Since basal risk has to be taken into account, these equations do not immediately show that they describe independent interactions. In situations where the increased risk to single risk factors is small, *independent* interactions can be described by an "approximate *additive* model" (Walter and Holford 1978), analogous to Eq. (3), describing the summation of effects

$$E_{A+B} = E_A + E_B,$$

in case of independent interactions where the effects E_A and E_B are small.

Similar to the findings of the interactions of chemical agents in pathophysiology (Chap. 8.1.3), the combined risk in epidemiology often exceeds independent effects of risk factors, e.g., with cardiovascular disease risks (see Pöch 1991b).

8.7 Herbicides

The combinations of pesticides, especially of herbicides, appear of great importance, because of the demands of farmers in modern food production and possible hazard for the crop and the ecology. Therefore, combinations of synergizing herbicides are sought (Gressel 1990), as well as combinations with insecticides, fungicides, and other agricultural chemicals, to save costs of energy and labor (Green and Bailey 1987). Thus, testing of herbicides in combination is necessary, and various methods have been employed (Bödeker et al. 1990).

Testing of a herbicide alone and in the presence of fixed doses of another herbicide or chemical has rarely been employed (see Gressel 1990), although it seems an economical approach for screening of herbicide combinations. DRCs of A and B, and of their mixture(s) have been tested by many authors (e.g., Altenburger et al. 1990, Streibig 1986).

A special problem of herbicide mixtures is *antagonism,* some of which appear to reflect independent rather than antagonistic interactions in isobolograms. Many do reflect antagonistic mechanisms (Green 1989). An example of *independent action* is the combination of atrazine and alachlor, described by Akobundu et al. (1975). There was a good correlation between observed combined effects and calculated effects of independent actions (Fig. 48d). Further, the observed effects match with independent DRCs (Pöch 1991b). The results of many combinations of herbicides, synergistic and antagonistic, can be explained by toxicokinetic rather than toxicodynamic mechanisms, i.e., due to enhanced or diminished uptake or metabolism (Gressel 1990).

8.8 Conclusions and discussion

This chapter has elucidated many different combinations in many different fields. From the examples given, many conclusions can be drawn with respect to methodology and evaluation.

8.8.1 Methods

All methods suggested and used in the evaluation of combined effects offer some advantages and disadvantages, at least in special cases. Taken together, the direct analysis of DRCs appears superior to the isobolographic evaluation. This appears to be true especially for conventional isobolograms which only rely on additivity as a qualitative and quantitative standard (see Chap. 9).

Dose-response studies: the simple approach
From the examples given, it can be seen that DRCs of A in the presence of one (!) well chosen fixed dose of B provide evidence which is in basic agreement with results of experiments with more than one fixed dose of B or with mixtures of A and B.

The simple approach is applicable to biochemistry, physiology, pathophysiology, pharmacology and toxicology. It is anticipated by the author that the effects of toxic agents and factors in combination will probably gain more and more importance in the future and will necessitate increasing numbers of experimental studies. A simple and cost saving procedure should better accomplish this goal than the widely used isobologram approach.

The new approach to apply a direct analysis of DRCs to combined effects of agents acting into the same direction is in analogy with antagonist studies. From a theoretical point of view, models can be invented for such studies with "synergists", and most importantly, the one of competitive synergism (Ariëns et al. 1956a) for a site-directed analysis. The mechanism of competition for a single binding site meaningfully explains the concept of additivity. This and other models, e.g., for the two-receptor one-transducer interaction (Undem and Adams 1988), are best applied to DRCs of one substance in the presence of fixed doses of another agent.

The practical aspects also favor, except in certain cases, testing a DRC of A in the absence and presence of a fixed dose of B, rather than a mixture of A plus B. This book for the first time provides comprehensive evidence for the advantages of the new approach compared with conventional procedures, notably, the isobologram approach. When experimental conditions are properly chosen, a study with a *single* fixed dose may provide conclusive evidence for the type of interaction and the magnitude of enhancement, where enhancement is present. In any case, the simple experimental approach to study the DRC of A in the absence and presence of one fixed dose of B appears useful, at least or especially for screening.

The only drawback of the new approach is that there is no software program yet available which "automatically" calculates curves for additive and independent interactions. However, fitting of experimental points to DRCs can be done by computer programs, which can also, and simply, be applied to the construction of the theoretical curves. Practical aspects, how this can be done, are given in Appendix B.

Dose-response curves: the elaborative approach
Where more than one dose of B is tested, the simple test can be expanded to incorporate several fixed doses of B but becomes more elaborate. Published experiments with a 5 × 5 factorial design or a similar protocol were reevaluated. Basically, the simple approach gives the same results as the elaborative procedure. Notable examples are the combined effects of ADP plus ADP-ribose, and ADP plus o-phenanthroline, shown in Figs. 36 and 37. The inhibitory effects of ADP on alcohol dehydrogenase appear additive with ADP-ribose at any given concentration of ADP-ribose (Fig. 36a), whereas the interaction with o-phenanthroline under the same experimental protocol at all doses of o-phenanthroline clearly deviates from additivity.

It should be emphasized that *deviations* from additivity may escape detection where the differences of observed and expected DRCs are small. The same applies to a comparison with independent DRCs. Recently, we studied the interaction of drugs with one mechanism of action in the presence of several fixed doses of a dualistic acting drug (Holzmann et al. 1992). Basically, identical interactions could be demonstrated with one fixed dose.

Dose-response curves of mixtures
An experimental example for the evaluation of a fixed dose-ratio combination, i.e., a mixture, has been given in Fig. 24. It shows that the mixture of two cholinesterase-inhibitors, malathion and parathion, did not significantly deviate from a calculated DRC for an additive mixture. This information could be obtained with parathion in the presence of 20 ppm malathion in Fig. 21. The other dose combinations used by Tammes (1964) also did not deviate significantly from additive with p-values between 0.08 and 0.91 (not shown).

Other procedures
This chapter also shows that other procedures, like time-course or single-dose studies may be appropriate to study combined effects, especially where DRCs cannot be established.

8.8.2 Comparison with model interactions

The interpretation of combined effects is based on a comparison of observed with calculated effects of additivity or independence. This cannot only be done in studies in which DRCs are established, but also in time course and single-dose studies. The comparison with additivity is not always feasible in other than DRC-studies but the comparison with independence is often possible.

Comparison with additivity
Mechanistically, additive combinations represent competitive interactions of drugs at one and the same binding site. Therefore, it is of interest to compare observed combined effects, preferably with DRCs, with calculated combinations of additive interactions, whenever we are interested in the site of action. Examples given in this chapter, other places of this book, and elsewhere (Pöch et al. 1990a) clearly support this view (e.g., Figs. 22, 36, and 37).

It should be kept in mind that *deviations* from additivity tell us about differences in the site of action, and that additive combinations may be due to coincidence. Only observed effects in combination which very well agree with additivity, in DRCs, prove a competitive interaction at the same molecular site (Fig. 36).

The examples of this chapter also show that many combinations can show the phenomenon of additivity, yet no competitive interaction can be expected. This is especially the case with drugs which possess steep and very steep DRCs, discussed in Chaps. 2 and 3. Where no interaction at the same molecular site can be expected, additivity has to be taken as a mere *phenomenon.* As a matter of fact, in most of the examples given in this book, and in almost all examples in chemotherapy and in toxicology, an inter-action at the same molecular site can reasonably be excluded. Numeri-cally, competitive interactions are rare, and must be rare. Hence, they are hardly of practical value in clinical pharmacology, chemotherapy, experi-mental and environmental toxicology, epidemiology, and in the field of pesticides.

So, there are some arguments to give up the additivity model, where we are not interested in a site-directed analysis or where we cannot expect chemical substances to interact competitively. Incidentally, physical factors cannot compete for a common binding site. Hence, we could still use the additivity model as a model for competitive interaction but not as the "gold standard" in the evaluation of all combined effects. Only where we are interested in the doses for effects in combination, does a comparison with additivity appear of value and appropriate for agents which are known, or strongly expected, to act at different sites (see Chap. 8.5). The weakness and the pitfalls of the isobologram approach based on the additivity con-cept are described and discussed in the following chapter.

From the above considerations it is also clear that additive combina-tions by competitive interaction require the *same* and not a similar action. The scheme for differentiation of combined effects used by some investiga-tors, e.g., by Chou and Talalay (1981), rests on a vague or misleading char-acterization of a "similar" action for agents acting at the same site (Hewlett and Plackett 1956).

Magnitude of enhancement/comparison with independence

Usually, the goal of studying effects in combination is to find out the mag-nitude of enhancement. This can be done in various ways, exemplified in this and in earlier chapters. A comparison with calculated effects of inde-pendent interactions appears of great value where we are interested whether the effect of one agent is affected (and in which direction) or not.

Most of the examples in clinical pharmacology (Chap. 8.3) and in later sections of this chapter show that the magnitude of combined effects often is comparable to or exceeds independent effects. From the considerations in Chaps. 2 and 3, we can conclude that greater than independent effects indicate potentiation. Hence, weak potentiation was usually observed in the mentioned sections, in contrast to physiology and pathophysiology (Chap. 8.1), where marked potentiations were seen (Figs. 38 and 39).

If future studies substantiate these findings, we can approximately *pre-dict* various combinations to conform to somewhat greater than indepen-dent effects – unless we have reason to expect a special type or mechanism of interaction, including pharmaco- and toxicokinetic interactions. Hence, we may in the future be able to predict combined effects by an appropriate model for several groups of agents on an empirical basis of a comparison of observed effects with independent effects in combination. Meanwhile, we may base predictions on the additivity model, where this appears pragmatically appropriate.

9 A new and critical view of isobolograms

Isobolograms have been mentioned throughout the foregoing chapters. Indeed, this graphical presentation of effects in combination is widely used since Loewe and Muischnek (1926) suggested it for the analysis of combined effects in pharmacology and toxicology. There are a number of objections to the general use of isobolograms, which in part have been touched in the foregoing chapters. However, many investigators consider the isobologram method as "the method of choice for the evaluation of possible interaction between biologically active agents" (e.g., Sühnel 1992). Given the popularity of these graphs, it appears important to look at isobolograms from a critical point of view.

The lack of theoretical DRCs of additive (and independent) interactions at that time may have prompted Loewe and Muischnek (1926) to publish this graphical procedure. In contrast to DRCs, isoboles show the *doses* of the constituents of a combination for a given effect. Theoretically, experimental isoboles can coincide with isoboles of dose-additive combinations or can deviate therefrom, indicating that in combination smaller or greater than additive doses produce the same effect as the individual compounds.

9.1 Inappropriate conclusions

Now, it is very interesting to note that Loewe and Muischnek (1926) defined such deviations in doses from additive as "synergistic" and "antagonistic", respectively. These as well as the terms "supra-" and "infra-additive" (over- and underadditive), used by Loewe (1953) and many others afterwards, are *effect-oriented terms*, despite the fact that they are based on doses for a certain effect. Hence, from the beginning, isoboles were taken to differentiate synergism (or potentiation) from antagonism. There is hardly any doubt that synergism and antagonism were meant to represent enhancement and diminution of effects, although Berenbaum (1989) has claimed to use these terms to describe phenomena. Even this author and others who followed his arguments (e.g., Sühnel 1990, 1992; Bödeker et al. 1990) ascribe the additive isobole to *zero-interaction* between A and B, a term that is interpretative, not descriptive. For describing the phenomenon of additivity, the term "zero-deviation" would be more appropriate.

One can argue that smaller or greater than additive doses for a given effect must logically mean that the same doses of A and B produce greater or smaller effects than an additive combination. This argument is certainly correct, but going beyond this point is a pitfall, i.e., when we generalize our conclusion in that sense that we consider overadditive and underadditive combinations to represent enhanced or diminished effects. This generalization is a pitfall, since we no longer compare assumed effects with effects of additive combinations but draw conclusions to effects without consideration of the relationship between changes in doses and changes in effect.

For instance, let us look at the situation termed "relative antagonism" (Loewe and Muischnek 1926, Loewe 1953, Calamari and Alabaster 1980), present if underadditive combinations do not require greater doses of A and B in combination than singly (Fig. 14a). It is clear that in the usual ED_{50}-isobologram smaller doses than the ED_{50}s of A and B will produce less than 50% effect when tested alone (unless this is the maximum effect produced by A and by B alone). Then, the isobole for these reduced doses indicate that 50% in combination is reached with doses of A and B which produce less than 50% each. This is enhancement in a broad sense, since the effect in combination is greater than the effect of the single constituents.

However, this is not generally meant when people talk about *enhancement* of A by B, although no common definition is yet available (see Chap. 1.4). There is good reason to express changes in effect with respect to independence in action (see Chap. 3.2) when B has an effect by itself. Then, an unaffected response conforms to independent effects in combination (see Chap. 2.3).

The relationship between the ED_{30} and the ED_{50} is dependent on the *steepness of the DRCs*. At a slope < 1.6, the ED_{30} is *less* than half of the ED_{50}, at a slope of > 1.6, it is *greater* than half of the ED_{50}. Therefore, the steepness of DRCs determines whether independent isoboles are overadditive, additive, or underadditive (Christensen and Chen 1985, Gessner 1988, Pöch et al. 1990c).

Although the steepness of DRCs is not shown in isobolograms, the relationship between independent and additive isobolograms (see Pöch et al. 1990a) is not difficult to understand on the basis of the above considerations. For simplicity, let us consider 50%-isobolograms. The individual effects, which in independent combinations produce 50% can easily be calculated, they roughly amount to 30%. Now, if the ED_{30}s of A and of B are *less* than half of the ED_{50}s, the independent isobole is below the additivity diagonal; hence, independent combinations are overadditive. However, if the ED_{30}s are *greater* than half of the ED_{50}s, the independent isobole is above the additivity line, hence this independent combination is underadditive.

From the above considerations, it is clear that additivity in isobolograms (and in DRCs) is not a good reference for enhancement in combination: An enhancement from single effects of 30% to 50% in combination may appear overadditive, additive, or underadditive.

If we look at isoboles primarily from the viewpoint of doses, we might see the zenith of an isobole of observed effects at doses reduced from 1 to

0.7, representing an underadditive isobole. However, the effects of A and B alone at the dose of $0.7 \times ED_{50}$ may be quite different – depending on the dose-response relationship. For instance, when A and B show a slope of 0.8 in DRCs, the individual effects of 0.7-fold ED_{50}s are 43%; when the slope equals 3, the effects of A and B alone are 26%; when the DRCs exhibit a slope of 5, the individual effects of A and B are 14% at the 0.7-fold ED_{50}. So, drugs at the same dose fractions show different magnitudes of enhancement with different steepness of DRCs. In the example given, at the 0.7-fold ED_{50}, this enhancement represents a change in effects from 43 to 50%, from 26 to 50%, and from 14 to 50%, respectively!

Any conclusions regarding the *magnitude of enhancement* based on additive isoboles are therefore inappropriate, including the "Variationsgrad" (Loewe and Muischnek 1926), the "combination index", CI (Chou and Talalay 1984), and the "synergism-antagonism parameter, α" (Greco 1987). This is not only true for isobolograms; the quantitative expression of deviations from additivity is also inappropriate for DRCs, including the "dose-factor of potentiation" suggested by the author in the first paper on combined effects (Pöch and Holzmann 1980/1981). These quantitative terms merely indicate the deviations from additivity of observed *doses* from additive doses.

We now can see that the conclusions, overadditive = enhanced, and underadditive = diminished *effects* in combination are obviously not justified. However, such conclusions are often found in the literature, albeit not always specified. On the other hand, to take independent effects as the *reference* is justified. Further, it appears more appropriate to compare observed effects with independent effects by direct analysis of *DRCs* than by isobolograms derived therefrom. We *see* the enhancement of effects in DRCs but not in isobolograms. Furthermore, in directly analyzing DRCs, we can save time and expenditure compared to an isobolographic analysis.

Why independence should be taken as reference for the separation and quantitation of observed combined effects has been discussed in Chap. 3. Briefly, independent effects of A represent effects which are unaffected by B, and vice versa, and are therefore comparable to unaltered effects in simple combinations.

9.2 Pitfalls

So far, we have evaluated and discussed isobolograms from the viewpoint of which conclusions are appropriate and which are not. One pitfall has been briefly mentioned. We will here discuss further why the statement "overadditive = enhanced, underadditive = diminished effects in combination" is not generally correct (see also Pöch and Reiffenstein 1992). Independent isoboles can either appear overadditive, additive, or underadditive; examples have been published recently (Pöch et al. 1990c). Hence, if we only rely on additivity, we cannot tell whether the effects in combination are enhanced, unaffected (independent) or diminished – unless we compare observed isoboles with independent isoboles (see Pöch et al. 1990c).

Following Loewe (1953), Chou and Talalay (1984) have considered additivity as the reference for diagnosing and quantitating synergism, "summation" and antagonism for mutually exclusive drugs. The latter bind to and compete for the same site (Cleland 1970, Wong 1975). Interestingly, Chou and Talalay (1984) concluded that Loewe's isobologram is "not valid for mutually nonexclusive drugs", i.e., for drugs which bind to different sites (Cleland 1975, Wong 1975). They have therefore calculated an isobole for mutually nonexclusive drugs for diagnosing and quantitating combined effects of mutually nonexclusive drugs. This isobole represents independent effects of inhibitors which show a slope = 1 of DRCs, and only for this condition (Gessner 1988, Syracuse and Greco 1986).

As a matter of fact, there is a discrepancy between "summation" isoboles for "nonexclusive drugs" (acting independently) with slope = 1 and those with slope values different from 1, already evident from the simple construction of (the zenith of) independent isoboles described recently (Pöch et al. 1990c).

Now, Chou and Chou (1987) take the "summation isobole", erroneously calculated for compounds with slope values different from 1, as the reference for diagnosing and quantitating synergism and antagonism, irrespective of the slope of DRCs. However, the higher the slope of DRCs, the greater the difference between correctly and incorrectly calculated independent isoboles! The procedure of Chou and Chou (1987) leads to the diagnosis "antagonism" whenever the observed isobole is above the summation isobole for nonexclusive drugs. Since the theoretical summation isobole represents an overadditive isobole, any experimental underadditive isobole is incorrectly judged to indicate "antagonism".

The Chou–Talalay procedure for the evaluation of combined effects (Chou and Talalay 1984) has gained popularity in the last years, not the least because a computer program is available (Chou and Chou 1987). The basic approach, however, is not clear or is incorrect. Briefly, additivity is used as reference for mutually exclusive, "independence" (incorrectly calculated) for mutually nonexclusive drugs. The differentiation between exclusive and nonexclusive drugs is based on the slopes of the linearized DRCs in the "median-effect plot" of the mixture A + B, compared to the components, A and B (Chou and Talalay 1984, Chou and Chou 1987). When the lines in this plot are all parallel, the drugs in combination are classified as exclusive drugs, if the line for the mixture (!) shows different slope values, the agents are treated as nonexclusive drugs.

Let us consider this point. We have discussed that agents acting at the same site (exclusive drugs) must show the same steepness in their individual DRCs, otherwise A and B could not be seen as dilution of one another. Also, Finney (1942) concluded that agents with "similar actions" (additive) show a common slope of the "regression lines" (DRCs). Differences in slope of the DRCs of A compared to B, indicate binding to different sites of action. Drugs with distinct sites of action are considered nonexclusive drugs by Chou and Talalay (1984), as well as by Cleland (1970) and Wong (1975). However, differences in the slope of A compared to B are *not* the basis of differentiation drawn by Chou and Talalay (1984) and Chou and Chou (1987). Though, it is not clear why not. As a matter of fact, it appears that these authors have followed the line of argument: (1) When the slope of the DRC of the mixture differs from the slope of the DRCs of the component, an action at the same site is excluded. Hence drugs A and B act at different sites. (2) When the slope is not different, then the drugs act at the same site. The latter conclusion is the reversal of the first conclusion, but it is unjustified, since drugs can also exhibit the same slope when they act at different sites. For instance, the central and peripheral analgesics, butorphanol and acetaminophen, tested alone and in a 1:10 and 1:125 mixture all showed parallel DRCs (Pircio et al. 1978). Interestingly, the combination was overadditive in the respective isobologram (Pircio et al. 1978). Also, agonists and antagonists acting at different receptors usually exhibit the same slope of around 1 (e.g., Ariëns et al. 1964a, 1979), the theoretical slope value for agents which actions conform to the law of mass action.

In the following chapter, the new approach described in this book, is compared with established and accepted procedures, not the least with the Chou–Talalay method, on the basis of concurring and controversial interpretation of combined effects.

10 Comparison of the new with the conventional approach

The first chapters of this book provided a new look at the theory for the evaluation of combined effects, the following chapter describes and discusses the practice of analysis and interpretation.

The approach to directly analyze combined effects by DRCs is new inasmuch as it relies on dose-response studies with a compound A in the presence of a *fixed dose* of B, and by comparing the DRCs of combined effects with *theoretical DRCs* for dose-additive and independent actions of A and B. It is the simplest DRC procedure, and therefore, requires less experiments than the fixed-dose ratio approach for the study of mixtures of A and B. The latter can be looked upon as the conventional approach for evaluation of combined effects (e.g., Altenburger et al. 1990, Bödeker et al. 1990, Gessner 1988, Hewlett and Plackett 1959, Sühnel 1992, Unkelbach and Wolf 1985a). With some exceptions (e.g., Unkelbach and Wolf 1985a, b), DRCs are not directly analyzed but "converted" to isobolograms prior to analysis. Some publications show the DRCs plus the isobolograms derived therefrom (e.g., Chou and Chou 1987; Chou and Talalay 1983, 1984; Pircio et al. 1987; Streibig 1986).

The utility of the new, direct analysis of DRCs of A with fixed doses of B has been demonstrated in previous chapters, and examples have been published recently (Pöch et al. 1990a, b; Holzmann et al. 1992). In the foregoing chapters experimental examples for time-course studies as well as single-dose studies have also been given, in which combined effects can also be compared with independent effects, emphasized in this book. However, DRCs usually provide the greatest amount of information. Therefore, DRCs are considered for this comparison with the conventional approach. Since the isobologram procedure has been discussed in Chap. 9 and a comparison of the new approach with the "mixture approach" has been given in Chap. 4.2, a modified isobologram method (Steel and Peckham 1979) and the method proposed by Chou (Chou and Chou 1987; Chou and Talalay 1983, 1984; Rideout and Chou 1991), will be the main topic of this chapter.

10.1 Experimental dose-response curves and isobolograms

In principle, we could reach the same conclusions from the analysis of DRCs and isobolograms as evident and illustrated by two examples, described in the following section. However, in reality, many controversial interpretations can be reached by different investigators. They will be discussed in Chap. 10.1.2.

10.1.1 Concurring interpretation

The DRCs of the combined effects of the enzyme-inhibitors ADP, ADP-ribose, and o-phenanthroline have been shown in Figs. 36 and 37. The combined effect of ADP plus ADP-ribose was strictly additive, whereas o-phenanthroline showed a marked overadditive interaction with ADP. These differences in the interaction between the inhibitors, ADP-ribose and o-phenanthroline, were at least as impressive as in the Yonetani–Theorell plot (Fig. 36). Isobolograms of the combined action of the inhibitors showed a strict additive behavior with ADP and ADP-ribose and a marked overadditive combination with ADP and o-phenanthroline (Chou and Chou 1987, Berenbaum 1989). So, it is clear that in these examples the different procedures for the evaluation led to the same result. We might add that the so-called median-effect plot also nicely differentiated between these two types of interaction (Chou and Talalay 1981, 1984; Chou and Chou 1987). However, there are some important points to consider.

The direct evaluation of the dose-response relationship by DRCs (Fig. 36a and b) or by the Yonetani–Theorell plot would have already provided the information "additive = competitive" with one of the five curves of ADP in the presence of ADP-ribose or o-phenanthroline, in addition to the control curve with ADP. The evaluation by Chou and Talalay (1981, 1984) involves two control curves and one curve of the mixture for one point of the mixture in the isobologram (Chou and Chou 1987).

In the experiments with ADP and o-phenanthroline, the combination appears overadditive in all procedures. However, further evidence for the type of interaction can be obtained from the nonlinear DRCs (Fig. 37) or from the linearized dose-response relationship in Fig. 36. While the isobologram analysis by Chou and Chou (1987) tells us that the combined effect of these inhibitors is markedly overadditive with a CI value at the 50% effect level (FA = 0.5) of 0.60, more information is available by comparison with independence. The DRCs as well as the Yonetani–Theorell plot show that the combined effects of ADP and o-phenanthroline exceed the effects of an independent combination.

In vivo experimental DRCs of an additive combination have been shown in Figs. 21 and 24 for the insecticidal action of two cholinesterase-inhibitors. Reevaluation of the respective tabulated data of Tammes (1964) showed no significant deviations from additivity, as nicely exemplified in Fig. 21, obtained with about 500 flies. The isobologram of Tammes (1964) shows four points of the mixture for 50% mortality rather close to the additivity diagonal, obtained from a total of more than 1000 flies.

Steepness of DRCs

As the examples of Figs. 36 and 21 show, the combined effects appear additive in DRCs and in isobolograms. These are examples in which the components of the combinations show the same steepness of DRCs. Of course, there is a relationship between DRCs and isobolograms, i.e., that DRCs which appear additive are also additive in isobolograms, at least if the components of a combination possess the *same steepness* in DRCs. Where the agents tested exhibit significantly *different slope values*, we still can construct an additive DRC of A in the presence of B, and vice versa, as exemplified (e.g., in Fig. 41 e and f). In this case, however, A and B do not bind to the same molecular site, hence additivity has to be interpreted as a phenomenon, only.

Some authors have also constructed additive *DRCs* for agents with different steepness, either as a single additive DRC of a mixture (e.g., Unkelbach and Wolf 1985b, Streibig 1986, Greco et al. 1990) or as an "envelope of additivity" (Teicher et al. 1991). The latter is an application of the "envelope of additivity" proposed by Steel and Peckham (1979) for isobolograms.

10.1.2 Different and controversial interpretation

Controversial interpretations arise when different investigators use different models and procedure for the analysis of combined effects. This section gives examples which lead to such controversies, the "envelope of additivity" and the nonconsideration and incorrect consideration of independence.

Envelope of additivity in isobolograms

Steel and Peckham (1979) considered that a different dose-response relationship for chemotherapeutic agents and radiation (!) does not permit the construction of a diagonal line in the isobologram for a combination of them. Interestingly, these authors consider chemotherapeutic agents and radiation to act independently (i.e., non-interactive) but refer to what "has often been called additive". The latter term is also used in the "envelopes of additivity" by Steel and Peckham (1979).

The concept of the envelope of additivity implies that drugs and radiation differ in their dose-response relationship with each other. As already pointed out, drugs with different steepness of DRCs cannot be regarded to conform to the "dilution model" of additivity. Further, "additive" combinations between drugs and physical factors (radiation) are necessarily excluded by this model from a mechanistic point of view.

Then it is inappropriate to compare combined effects with the envelope of additivity if we want to check whether the agents in combination act at the same binding site. If we are interested whether or not agents in combination exhibit enhanced effects, i.e., are potentiated, we have to compare the combined effects with the effects of independently acting agents or factors. In contrast to a mechanistic additive interaction, independent effects can occur between physical factors as well as with chemical compounds. What an envelope of additivity of DRCs as well as isobolograms tells us is that the agents or factors in combination exhibit different steepness of DRCs. The wider this envelope, the greater the differences in the steepness of DRCs.

The important point in the controversy between the new analysis of DRCs and the modified isobologram procedure of Steel and Peckham (1979) actually is not the envelope of additivity but the fact that Steel and Peckham (1979) interpret their results with respect to additivity and along the conventional line (see Chap. 9).

The concept and method of Steel (Steel 1979, Steel and Peckham 1979) has been adapted for drug–drug interactions, e.g., by Tsai et al. (1989). The data of that paper have been reevaluated in this book (Figs. 34c, d and 35c, d) and were found to be comparable with an independent interaction of cisplatin and etoposide on cell survival. It appears that the interpretation of isobolograms with respect to the envelope of additivity leads to the same conclusions as the reevaluation shown in Figs. 34 and 35. However, the claim of lack of synergism by Tsai et al. (1989) rests on the comparison with additivity, whereas the reevaluation shows an independent response in combination, i.e., lack of potentiation.

Nonconsideration and incorrect calculation of independence

Different interpretations of combined effects expressed by DRCs and by isobolograms are mainly due to consideration and nonconsideration of independent effects. It has been shown recently that independent combinations can also, and easily, be constructed in isobolograms (Pöch et al. 1990c). There is, as one might expect, good agreement between DRCs and isobolograms with respect to independent effects (see Pöch et al. 1990c).

Differences in the interpretation of combined effects are preprogrammed when one evaluation compares combined effects with calculated independent effects, and the other does not, or is based on incorrectly estimated independent combinations. Thereby, one can reach quantitatively and qualitatively different conclusions with respect to potentiation (synergism) and antagonism. As pointed out before, independent effects are of greatest importance as a reference for comparison with observed effects in combination, and for the differentiation of potentiation (synergism) and antagonism (see Chap. 3).

Differences in interpretation are particularly striking when substances with very steep DRCs are evaluated in combination. Since in this situation independent combinations are underadditive, thus, they are considered by many authors as "antagonism", especially by those who rely on isobolograms (see Fig. 14).

Another reason for controversies in the interpretation of combined effects arises from the incorrect calculation of independence by Chou and

Talalay (1984) as stated by Gessner (1988) and Syracuse and Greco (1986). The method of Chou and Talalay involves the "median effect plot", a presentation of linearized DRCs of A, B, and a mixture of A plus B. This plot shows the steepness of the DRCs, from which the arbitrary diagnosis of A being "mutually exclusive" or "mutually nonexclusive" with B is derived. This diagnosis has been discussed critically in Chap. 9. From the basis of isobolograms, Chou and Talalay quantitate the deviation from additivity and the incorrectly estimated independence by the Combination Index (CI). CI < 1 is taken as synergism, CI > 1 as antagonism.

A comparison of the new approach with DRCs of A in the absence and presence of a *fixed dose* of B with the procedure of Chou and Talalay cannot be given because the latter requires testing of *mixtures*. Nevertheless, it appeared of interest to compare non-linear fitting of DRCs of the mixture with the linear fitting used by these authors, expressed by DRCs (Chou and Chou 1987; Chou and Talalay 1983, 1984). Two examples show the inferiority of DRCs obtained by linearized fitting as well as the controversies which can arise from the incorrect calculation of the isobole for independent interactions, termed mutually nonexclusive interaction, with respect to correct determination of independent isoboles (Pöch et al. 1990c).

Interaction of anticancer agents

Hofmann et al. (1989) published DRCs of two anticancer drugs, cisplatin, BM41440, and their 1:10 mixture. From their figure 5A DRCs and isobolograms were obtained by two procedures. In Fig. 49a and b, the log-dose response curves were obtained from fitting the experimental points by nonlinear curve fitting (ALLFIT), and the corresponding isobologram for 50% effects. Figure 49c and d shows linear-dose response curves, obtained by fitting the points to the "median effect plot" (Chou and Talalay 1984, Chou and Chou 1987) and the corresponding isobologram.

Curve-fitting by ALLFIT resulted in the DRCs of Fig. 49a, the experimental points were very closely fitted to the DRCs with maxima of the single components definitely below the possible maximum. The slopes of cisplatin and of BM41440 were estimated to equal 3.3 and 1.5, respectively.

The DRCs in Fig. 49c show a poor fit of the experimental points of BM 41440 to its curve, also the DRC of cisplatin clearly exceeds the low maximum in Fig. 49a. The DRCs in Fig. 49c show considerably lower slope values, i.e., 1.1 for cisplatin, and 0.8 for BM 41440. Fitting points to transformed straight lines in general is not as good as that obtained from nonlinear fitting (Motulsky and Ransnas 1987, Syracuse and Greco 1986). Furthermore, linear fitting assumes an effect range between < 1 and > 99%, i.e., that the maximum of 100% is reached.

Due to the different slopes of the DRCs, the calculated isoboles for independent actions differ (compare Fig. 49b with Fig. 49d). In any case, however, the combined effects appear more pronounced than expected for independent combinations, i.e., lower doses than for independent

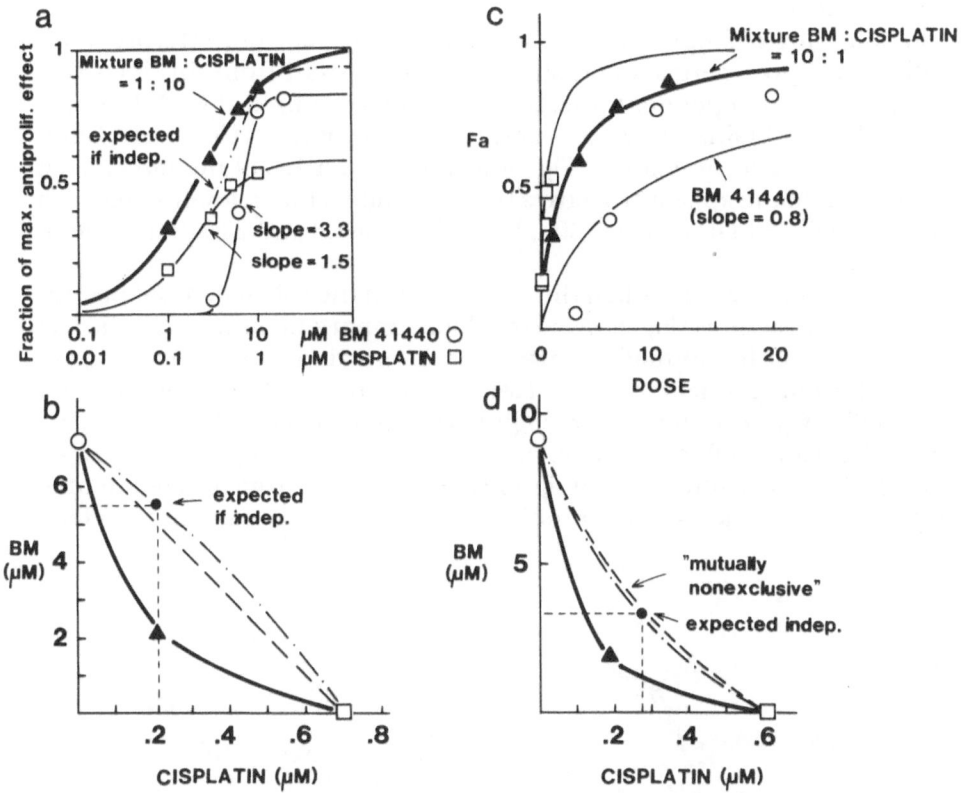

Fig. 49. Comparison of DRCs of mixtures with corresponding isobolograms, exemplified by the antiproliferative effects of cisplatin (□) and BM 41440 alone (○) and as a mixture (▲) in a fixed dose ratio (data of Hofmann et al. 1989). **a, b** DRCs fitted by ALLFIT (**a**) and isobologram derived from it (**b**). **c, d** DRCs fitted by the software program DOSE of Chou and Chou (1987) with doses on a linear scale (**c**) and corresponding isobologram, as obtained by that program (**d**). **b, d** The isobolograms also indicate expected independent combinations (broken dotted line), **d** the isobole for "mutually nonexclusive" agents (broken line), **b** and additive (broken line) interaction

interactions are needed to produce 50% effect. It is interesting that there is hardly any difference between the isobole for independent and the one for mutually nonexclusive interaction in Fig. 49d. The explanation is that the isobologram was constructed for two agents with a slope around 1. At higher or lower slope values, "mutually nonexclusive" isoboles differ from isoboles of independent action, as can be seen from Fig. 50d. Others have also stressed the point that the construction of an isobole for "mutually nonexclusive agents" by Chou and Talalay (1984) does not consider actual slope values (Gessner 1988, Syracuse and Greco 1986, Berenbaum 1989). Independent isoboles constructed as described in this book are in agreement with independent isoboles of Gessner (1988) as well as Christensen and Chen (1985).

Interaction of insecticides

The following example shows that a *qualitatively* different interpretation with respect to synergism and antagonism can result from this incorrect estimation of independent interactions. In this example also, differences in the DRCs obtained by ALLFIT- and the DOSE-program (Chou and Chou 1987) are seen for lindane in Fig. 50a and c. The DRC of the insecticidal action of lindane appears to have a slope of 6 and a much lower maximum than the DRC of DDT in Fig. 50a, but a slope of 1.8 and a high maximum in Fig. 50c.

These differences explain the differences in the calculated isoboles for independent actions (Fig. 50b and d). More important, however, is the fact that the isobole for mutually nonexclusive interaction is markedly different from the independent isobole. Hence, the effects of the combination markedly (Fig. 50a and b) and slightly (Fig. 50d) exceed those expected for independent effects. Hence, the combined effects appear potentiated in Fig. 50a and b, although not significantly different from independent in Fig. 50a. Also, the isobolographic analysis indicates combined effects

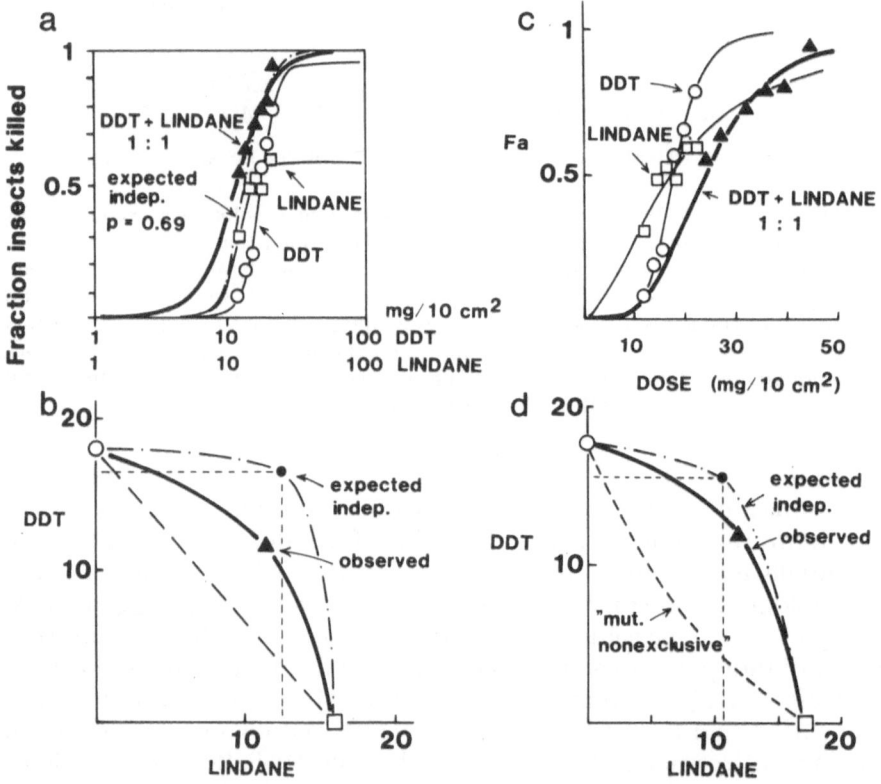

Fig. 50. Analogous to Fig. 49 but with the insecticidal effects of DDT and lindane, derived from the data of Hewlett and Plackett (1950)

greater than independent. Clearly, the results in Fig. 50a and b do not indicate antagonism on a mechanistic basis. However, the combination in Fig. 50d would have been considered as "antagonism" by Chou and Talalay (1984), since the experimental isobole is above the one for mutually non-exclusive interactions. The distinction between "mutually exclusive" and "mutually nonexclusive" drugs is based by Chou and Talalay (1984) on the slopes of the DRCs, graphically expressed by the "median-effect plot", a Hill-type plot (not shown). The weakness of this approach has been discussed in the foregoing chapter.

Many researchers have so far studied fixed-ratio combinations, i.e., mixtures of drugs with constant ratios between the doses of A and B, following the method of Chou and Talalay (1984). The computerized evaluation provides DRCs on a linear-dose scale (see Figs. 49c and 50c), median-effect plots, tables as well as graphs of the Combination Index (CI), and isobolograms. Despite the convenience to analyze combined effects in a one-step procedure by a computer program, it has serious drawbacks, limitations, and contains a fundamental error, which was discussed above.

10.2 Evaluation of combined effects in the future

Although our approach, so far, is not as convenient as the procedure of Chou and Chou (1987), it appears more straightforward and it provides much more information. It is hoped that this new approach will be used by those who are interested in a correct and meaningful evaluation of combined effects, and that it will aid our understanding of effects in combination. This, in turn, should provide a better appraisal of combined effects and their prediction with less experimental expenditure. It is also hoped that the direct analysis by DRCs of combined effects (Pöch 1991a) will again bring scientists together and enable them to understand the analysis of effects in combination carried out by colleagues.

It would even help the communication among various investigators if those who want to use the isobologram technique in the future would appreciate the limitations of the additivity model for the qualitative and quantitative analysis. Also, it is suggested that those researchers complete their isobolograms by addition of independent isoboles, which can easily be constructed (Pöch et al. 1990c).

The consideration of independent effects in combination appears one of the most important points, regardless of whether they are calculated as effects, e.g., in DRCs, or as doses expected for independent interactions in isobolograms. This could in the long run end the continuing debate on potentiation or synergism on the one hand and antagonism on the other, and the quantitation of combined effects.

Epilog

This book described and discussed the fundaments of a modern, meaningful and practical evaluation of combined effects, and hence, its theory and practice. The reader is expected to understand the current dilemma of disagreement in the analysis and interpretation of effects in combination after a careful and unbiased study of this book. Possibly a few points still remain unclear.

Two models of types of interactions have been described and critically looked upon, *additivity* and *independence*. The universal use of additivity has been challenged, probably for the first time, on the basis of modern understanding of the action of drugs and other chemicals. The vague term of a *similar action* is still the basis of additivity for many authors. However, there is evidence that agents truly acting alike, as in a sham combination, have to bind to a single *molecular site*, rather than to a "primary site of action" or the like. On the basis of this understanding, two important conclusions can be drawn.

Firstly, the additivity concept allows appropriate investigations aimed at the molecular site of action of chemical substances. There, additivity corresponds to a special type of competitive interaction. This model cannot be applied to physical factors appropriately, since the latter do not bind to certain sites.

Secondly, the additivity model is neither essential nor appropriate in all other situations, i.e., in the great majority of all investigations. Where the doses should be indicated, we do not need the model either; we can use the "model-free" FIC- and FEC-values, respectively, to clearly describe a combination.

A comparison of experimental combined effects with the effects predicted by the *independence* model has a much greater impact than the additivity model. The concept of this model can simply be described as that type of joint action in which the action of one compound or factor is not affected by other compounds, i.e., neither enhanced nor diminished. Again, important conclusions can be drawn.

First, the model does not require (complete) DRCs; it can also be applied to time-course studies and to single-dose studies. Theoretical DRCs of independent actions of A in the presence of a fixed dose of B can easily be constructed since they share the ED_{50} and slope with the DRC of A. There is no situation where a comparison with independence is not appropriate. A comparison with independent actions is not done because we expect experimental combined effects to precisely conform to it but because independent effects represent unaffected effects.

Hence, second, independent effects can appropriately be taken as the reference for differentiation and quantitation of *potentiated* or *antagonized* actions. Such an approach is in perfect agreement with the differentiation and characterization of potentiation and antagonism in situations in which B has no effect on its own.

Third, empirical investigations already show that the combined effects of compounds which bind to different molecular sites, or environmental factors, usually correspond to or exceed calculated effects of independent (inter)actions. This finding allows a cautious prediction of combined effects on the basis of the effects of the single compounds or factors.

Regardless of whether or not additive and/or independent effects are considered for evaluation of combined effects, dose-response studies can directly be analyzed, and costly isobolograms are needless. Additive DRCs, i.e., DRCs of competitive interactions, can rather easily be constructed by following the practical guidelines of this book. Where comparisons with independent effects are to be done only, time-course and single-dose studies can also be used.

Altogether, the modern approach described in this book offers a solution to the problem of how to analyze combined effects from the theoretical and practical point of view. It is suggested that a comparison of experimental with independent effects is done in order to characterize the type of effect in combination, i.e., (lack of) potentiation/synergism and antagonism. Only where the requirements for a competitive interaction are given, a comparison with additivity appears appropriate. Where possible, the evaluation should preferentially be based on DRCs.

Appendix A
Glossary of terms and abbreviations

Definition and characterization uses "A" and "B" etc. to represent chemical agents, or sometimes, physical factors which are studied in combination. "DRC" is the abbreviation for dose-response curve. In italic, cross-references.

Additive Differently used term, either describing a combination which corresponds to the *additivity* model (*dose-additive*) or a combined effect corresponding to the addition of single effects (*effect-additive*). Also, an effect which is increased by another active agent. In the sense of *non-interactive*, the term "additive" is also used to indicate an *independent* action of drugs etc. in the literature.

Additivity Model of combined action which describes the effect of A in the presence of B as if A in the presence of itself had been investigated (*sham-combination*). Thereby, A behaves like a dilution of B or vice versa. The combined effect can be calculated on the basis of the equieffective doses of the drugs.

Additivity envelope The region of *additivity* instead of the diagonal line in an *isobologram*, or instead of the additive DRC. The additivity envelope was invented and constructed for the combined effect of chemotherapeutic agents and radiation, with differences in the shape or steepness of DRCs (Steel 1979), neglecting that the concept of *additivity* is based on doses of chemical agents.

Antagonism Mechanistic or descriptive term. Broadly, working against each other. The result of *antagonistic* interaction of drugs, where one drug counteracts the action of another drug (mechanistic) or less than "expected". Often describing combined effects with respect to *additivity* or, less frequently, to *independence*.

Antagonistic Acting together to produce *antagonism*; the interaction of A and B where the effects of A are reduced, counteracted, or diminished by B or vice versa (mechanistically). As a descriptive term to indicate a combined effect which is less than "expected", usually *underadditive* or *subadditive*.

Antagonize The action in which one agent diminishes the effect of another agent resulting in *antagonism*.

Binding site The site to which molecules bind due to chemical forces between corresponding molecular groups. Binding sites are known as *receptors* for transmitters or hormones but are also found on other structures, like enzymes or ion-channels, as sites for various agents, the action of which is considered *specific.*

Chou–Talalay method Method for evaluation of combined effects employing linearized DRCs (median effect plots) and isobolographic analysis, quantitatively expressed by the *Combination Index* (Chou and Talalay 1984). It is based on differentiation of *mutually exclusive* and *mutually non-exclusive* drugs, the *isobole* of the latter is wrongly estimated for drugs with DRC *slope* > 1.

CI Abbreviation for *Combination Index* and *Coefficient of Interaction.*

Coefficient of Interaction (CI) Ratio of expected sum of fractional doses of drugs if they are additive (i.e., 1) to the obtained sum of fractional doses; CI > 1 = overadditive, CI < 1 = underadditive. Interpretation by Kissin et al. (1990): CI > 1 = synergism, CI < 1 = antagonism. The reciprocal of the *Combination Index.*

Combination Index (CI) The quotient of equieffective doses observed and expected (see Chap. 3.6.1) as a quantitative measure of deviation from additivity (*mutually exclusive*) and independence (*mutually non-exclusive*), respectively, if the latter are derived from DRCs with a slope = 1. Descriptive interpretation according to Chou and Talalay (1984), which is inappropriate as a measure of enhancement or diminution: CI = 1 (*additive*), CI < 1 (*synergistic*), CI > 1 (*antagonistic*).

Competitive (interaction) A mechanistic term for the competition of chemical molecules for a common *binding site*, e.g., a *receptor*. Depending on the *intrinsic activity* of the competing agents, the interaction can be characterized as competitive antagonism, competitive dualism, or competitive synergism. The latter term corresponds to *dose-additive* interaction (see Chap. 2.1.4).

Complex similar action (Plackett and Hewlett 1952) "Interactive *joint action*", in contrast to *simple similar* action.

Concentration addition Synonym of dose addition or *dose-additive.*

Dissimilar action Different sites of "primary action" of two drugs. According to Plackett and Hewlett (1952), *independent* if *non-interactive,* dependent if *interactive;* of historical interest.

Dose-additive *Additive* in the sense of (a combined effect explained by the) addition of doses, e.g., in the *isobologram.*

Dose-response curve (DRC) Graphical expression of the dose response to a drug or chemical, from which the following characteristics can be obtained, E_{min}, E_{max}, ED_{50}, and *slope.*

ED_{50}, EC_{50} The effective dose (D) or concentration (C) which exhibits 50% of a maximum response. In the analysis of DRCs in pharmacology, it is usually defined as the halfmaximum effective dose or concentration, irrespective of whether or not the ED_{50} or EC_{50} actually corresponds to the 50% effect level of the parameter. In the *isobolographic* analysis, the ED_{50} refers to the doses (in combination) by which the 50% effect level is reached, irrespective of E_{max} and E_{min} in DRCs.

Effect-additive Summation of effects. Effect-additive combinations roughly correspond to *independent* combined effects when A and B exhibit very small effects, otherwise it is a pure descriptive (and often confusing) term. Mix-up with *dose-additive* is likely to occur, when termed "*additive*".

Effect-multiplication The combined effect is the product of the effects of its constituents, e.g., when the effects are expressed by the fraction-of-control (FC) values: $FC_{A+B} = FC_A \times FC_B$. Special case of *independent* action which also describes a *dose-additive* combination of agents if they exhibit exponential DRCs (see Chap. 2, Fig. 8).

E_{max} The maximum of a drug effect (in DRCs). May be less than the possible maximum.

E_{min} The minimum of a drug effect (in DRCs).

Expected effect The calculated effect of either a *dose-additive* or an *independent* combination, depending on the expectation or view of the investigator.

FIC Abbreviation for "fraction of equieffective inhibitory concentration(s)", e.g., of A in the presence of B.

FIC Index (Elion et al. 1954) The sum of *FIC*-values in combination as a quantitative expression of over- or underadditive combinations, *synergism* and *antagonism*.

Fractional-product method Method for calculation of *independent* inhibitory effects in combination, expressed by values of fraction of control. Invented by Webb (1963) who described the combined response as the product of the individual fractions of control, hence in correpondence to the *multiplicative* model (see Eq. (4) in Chap. 2.3.1). Also known as the Webb method.

Heterergic A combination or drug pair which is not *homergic*, e.g., A is active, B is inactive.

Heteroadditive (Frei 1913) corresponds to *effect-additive*.

Homergic A combination of components producing phenomenologically the same overt effect.

Independence The important model of *independent* action of two, or more than two, agents or factors, based on the effects of them.

Independent (combination, action, effect) Unaffected response of A in the presence of B, as well as vice versa, i.e., unaltered net effects. The total effects of independent actions of A and B (and more than two drugs or factors) can be calculated by numerous equations (see Chap. 2.3.1). The mechanistic interpretation of combined effects corresponding to independent actions is that A acts independently of B or other agents or factors, and vice versa.

Infraadditive Synonym of *overadditive*.

Interaction (1) The (result of any) action between two or more agents or factors (this book). (2) An action in combination which deviates from an *expected effect*.

Interactive Descriptive term to indicate that the magnitude of effects in combination deviate from an *expected effect* (see *non-interactive*).

Isoadditivity (Frei 1913) corresponds to *additivity* in the sense of *dose-additive*.

Isobole (Isobol) A line in an *isobologram* indicating equal effects in combination, obtained by fitting experimental data points (fractions of doses), or by theory.

Isobologram A widely used graph (isobolograph) in which the doses of A and of B are plotted to obtain a specified effect, usually 50 %, tested alone as well as in combination (see Fig. 4).

Isobolograph Synonym of *isobologram.*

Joint action Effect of A and B (etc.) acting together, i.e., in combination, e.g., *independent* joint action, and *similar joint action* (Bliss 1939).

Multiplicative Descriptive term for combined effects which mathematically conform to *effect-multiplication*, e.g., in survival.

Mutually exclusive Two (or more than two) agents which share the same site of molecular action, which bind to the same site, hence compete for it when tested in combination, analogous to the competition between substrate and competitive inhibitor for binding at the substrate binding site. Also used in a less precisely defined situation in which A and B behave "similar" and exhibit an additive combination in the isobologram (Chou and Talalay 1984).

Mutually non-exclusive (1) The negation of *mutually exclusive* (binding) if agents bind to distinct molecular sites and (2) act independently of each other in a combination (Yonetani and Theorell 1964).

Non-interactive Confusing term indicating that observed or assumed combined effects correspond to or match with the *expected effects* (see *interactive*).

Non-specific Historically, a drug or other chemical which does not act like a *specific* agent but acts due to physical properties, like lipid solubility, unrelated to the three-dimensional position of chemical groups in the molecule. Nowadays, a drug or other chemical which is likely to bind to many hydrophobic regions on proteins at one and the same time, due to high lipid solubility.

Overadditive (1) A combination which is more pronounced than in a dose-additive combination in DRCs. (2) The doses of agents in combination are smaller than expected for dose-additive combinations, e.g., in isobolograms.

Potentiate The action in which one agent enhances the effect of another agent resulting in *potentiation.*

Potentiated An enhanced action. Many different definitions and uses of the underlying phenomenon and/or mechanism of *potentiation.*

Potentiation The phenomenon of enhanced action, clearly characterized by a reduction of the ED_{50} (and or an increase in E_{max}) of A by a fixed dose of B (which lacks an effect by its own). In the opinion of the author, also appropriate to describe enhanced effects in conditions where B alone exhibits an effect. Often used as synonym of *synergism.*

Potentiator A compound or factor which *potentiates* the action of (an)other compound(s) or factor(s).

Receptor In pharmacology and toxicology: an area on a (protein) molecule, a *binding site* for endogenous transmitters or hormones and exogenous agents which often share special molecular features with endogenous substances.

Relative antagonism According to Loewe and Muischnek (1926) an underadditive combination in which the doses in combination are not greater than those of the singly applied agents (e.g., Fig. 14d).

Response addition According to Dawson (1991): "An effect greater than that for either compound alone, but less than that for *concentration addition*". Corresponding to *relative antagonism*.

Response surface Three-dimensional (3-D) expression of dose response of A and B in combination. Quite popular nowadays because of the availability of computer programs. From the 3-D graphs, two-dimensional *DRCs* and *isobolograms* can be derived.

Response-surface methods Methods based on the *response surface*, e.g., the Universal Response Surface Approach (Greco 1987)

Sham-combination A combination between A and A, e.g., the DRC of A in the presence of a fixed dose of itself.

Similar action of A and B (1) Phenomenologically similar, (2) mechanistically the same action (at the same site!), e.g., *similar joint action*. Subdivided by Plackett and Hewlett (1952) in "simple similar" and "complex similar"; of historical interest. The term "similar action" was also used by Finney (1942) to mathematically describe *dose-additive* combinations.

Similar independent action Describes *independent* effects in combination (Bliss 1939).

Similar joint action This term was used by Bliss (1939) to describe the combined effects of two poisons, "acting upon the same system of receptors", corresponding to the *additivity* model.

Simple similar action (Plackett and Hewlett 1952) Corresponds to *additive* action, considered *non-interactive*. Competitive interaction was considered as a form of *complex similar* action; of historical interest.

Slope A measure of the steepness of a curve or line, usually corresponding to the Hill coefficient n_H. Slope = 1, < 1, > 1, depending on the increase in response with increase in dose.

Specific A drug or substance which binds to a (single) special *binding site* because of the respective molecular groups. This term is also often used instead of "selective".

Steepness of DRCs Plotting different DRCs at the same x- and y-axis results in different steepness. The steeper the DRC, the greater the increase or decrease in effect with the same increase or decrease in the dose, mathematically expressed by the *slope* of DRCs.

Subadditive Synonym of *underadditive*.

Summation (Webb 1963) Another term for independent effects, applied to inhibitory effects i of inhibitors 1 and 2: $i_{1,2} = i_1 + i_2 - i_1 i_2$. Reference for differentiating *synergism* and *antagonism*.

Supersensitivity Increased sensitivity, expressed by the ED_{50} and/or E_{max}; synonym of *potentiation*.

Supraadditive Synonym of *overadditive*.

Synergism The result of a *synergistic* interaction. A combined effect which exceeds the *expected effect.* Many different definitions.

Synergistic Acting to produce *synergism.*

Synergize The action in which one agent enhances the effect of another agent resulting in *synergism.*

Synergy Synonym of *synergism.*

Toxic units (TU) The dose which produces 50% response (ED_{50} as defined in *isobolograms*) of each compound tested in combination is defined as 1.0 TU. Toxic units in combination < 1 = *overadditive,* > 1 = *underadditive.*

Underadditive (1) A combined effect which is less pronounced than in an expected dose-additive combination in DRCs. (2) The doses of agents in combination are higher than expected for dose-additive combinations, e.g., in isobolograms.

Webb method See *fractional product method.*

Zero-interaction The result of and the magnitude of combined effects which is not different from that of an "expected effect" because A and B are considered *non-interactive.*

Appendix B
Guide to practical work: exercise examples

This practical section provides examples of DRCs of a compound A alone and in the presence of a fixed dose of another agent, B. This approach has been outlined in Chap. 4, and examples have been given recently (Pöch et al. 1990a, b) and in various parts of this book, with a linear dose scale (Fig. 15) or with a log-dose scale (e.g., Figs. 21 and 22). Also, the statistical analysis by the chi-square goodness-of-fit statistics was described in Chap. 4.1.2 and illustrated in Fig. 20 for a quantal (frequency) and a quantitative (graded) response.

There are several practical aspects, i.e., the construction of theoretical curves, the statistical evaluation, and the prediction of effects in combination. Two examples describe in detail how theoretical curves can be constructed either without (Example 1) or with computer-aided curve fitting (Example 2), and how the statistical comparison of an observed with "expected" response can be done, i.e., for a frequency (Example 1) and a graded response (Example 2), respectively. Other examples are provided as tabulated data (Tables 16 and 17) with reference to figures in this book. So, the latter examples are given for those who want to practise.

Construction of theoretical DRCs

Although the reader should be able to calculate and construct theoretical curves for additive and independent combinations after having studied Chap. 4, it nevertheless appeared appropriate and of value to provide examples for practical examination for two reasons, (1) as practical guide, also for technical assistants and other non-specialists, (2) to help avoiding mistakes and to enable the reader to more quickly and efficiently use the program ALLFIT.

DRCs can be evaluated with or without using a curve-fitting program. Therefore, one example each is given for self examination. When ALLFIT is used, the reader can use tables to find the ED_{50} and the slope of the respective additive DRC, provided the steepness of the DRC of A shows a slope not greater than 2.4. This simplifies the construction of theoretical curves compared to the one outlined recently (Pöch et al. 1990a). Since the calculation of independent effects is quite simple, no tables are provided.

Example 1. Frequency response
Evaluation without ALLFIT

We first look at the observed frequency (obs. frequency) of insect lethality
exposed to parathion alone and to 20 ppm malathion alone (Tammes
1964), from which we want to calculate the respective DRCs for an additive
and an independent combination, respectively, without ALLFIT (because
the slope of the DRC of parathion is 2.83).

ppm parathion	0	0.625	1.25	2.5	5
% obs. frequency	0	1	8	54	83

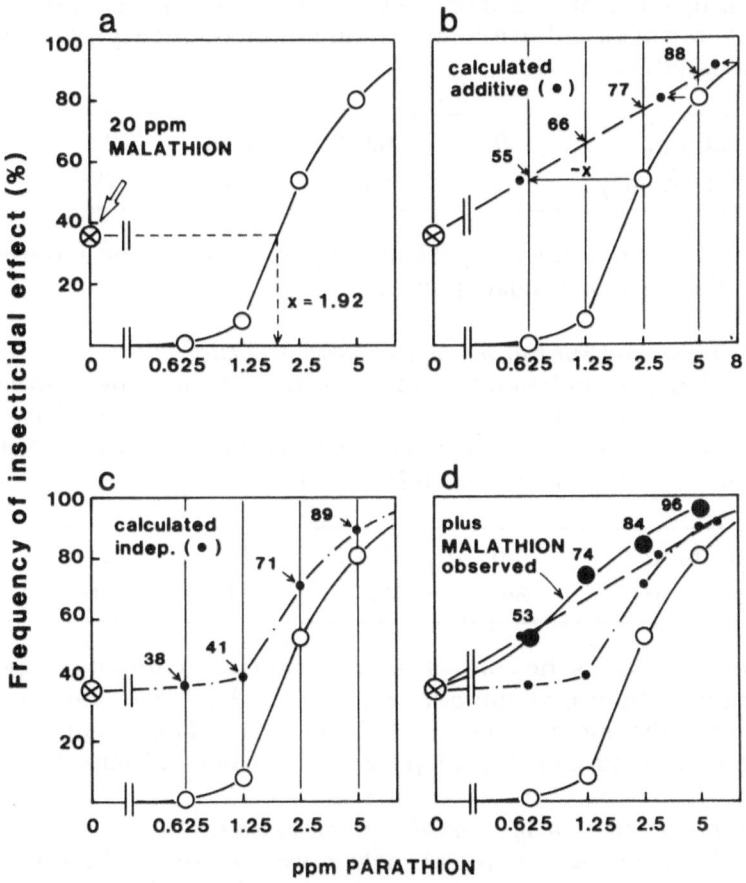

Fig. 51. Illustration of construction of theoretical DRCs: Example 1. **a** First step: estimation
of equieffective dose x. **b** Second step: calculation of points of the additive DRC by subtract-
ing x from doses of A (parathion) (○), and estimation of response at the doses of A applied
as indicated. **c** Calculation of points of the DRC of independent effects at the doses of A
applied. **d** Third step: comparison of observed response (●) with calculated response
(additive, broken line; independent, broken dotted line)

The observed frequency at 20 ppm malathion was 36%

Figure 51a show dose-frequency curves of parathion, fitted by eye. The effect of 20 ppm malathion corresponds to 1.92 ppm parathion, which we designate by x.

Calculation and construction of additive DRC

Having estimated the equieffective dose x, we can now calculate a few points for the additive dose-frequency curve, since we know that in additive combinations malathion must behave like x of parathion. Hence, the same frequency is expected in additive combinations at the applied doses of parathion minus x: Fig. 51b illustrates three points of the additive DRC, estimated at the experimental doses of 2.5 and 5 ppm, and at the extrapolated dose of 8 ppm parathion. The effect at these doses is expected at doses minus x, i.e., at 0.58, 3.08, and 6.08. Connecting these points shows the frequency of an additive combination (addit. frequency) at the applied doses of parathion:

ppm parathion	0	0.625	1.25	2.5	5
% addit. frequency	36	55	66	77	88

Figure 51b shows the control curve of parathion and the additive DRC, fitted to the above listed data points.

Calculation and construction of the independent DRC

The calculation of independent effects can easily be done following the equation $E_{A+B} = E_A + E_B - (E_A \times E_B)$, when we express the effects by the fraction of maximum frequency. For instance, the independent effect at 1.25 ppm = 0.08 + 0.36 − (0.08 × 0.36) = 0.41 = 41%.

ppm parathion	0	0.625	1.25	2.5	5
% indep. frequency	36	38	41	71	89

Figure 51c shows the control curve and the curve for an independent combination. Note that the additive DRC in Fig. 51b exhibits greater effects than the independent DRC in Fig. 51c. This is correct, and a characteristic feature of very steep DRCs, described in Chap. 3.

Statistical comparison of observed with calculated DRCs

If we are interested in a statistical analysis, we can use the chi-square goodness-of-fit statistics. Note, that in frequency studies we have to express the frequency by the observed and expected numbers of reactors. We therefore have to calculate the number of reactors by multiplying the fraction of killed insects with the total number used in each dose group (see Pöch et al. 1990b). We have to do this for the observed frequencies in combination, the additive, and the independent frequencies (Table 10).

Table 10

Parathion (ppm)	Observed % frequ.	no. react.	Additive % frequ.	no. react.	Independent % frequ.	no. react.
0	36		36		36	
0.625	53	26.5	55	27.5	38	19
1.25	74	37	66	33	41	20.5
2.5	86	42	77	38.5	71	35.5
5	95	47.5	88	44	89	44.5

We now compare the observed number of reactors with the expected number of reactors in the χ^2 goodness-of-fit test, using a suitable software program, e.g., STATGRAPHICS (Statistical Graphics Corp., STSC, Rockville, Md., USA).

Comparison of observed with additive (Table 10):
$\chi^2 = 1.12$ with 3 degrees of freedom (d.f.); p = 0.77.

Comparison of observed with independent (Table 10):
$\chi^2 = 17.63$ with 3 d.f.; p = $5.23 \times 10^{-4} = 0.000523$.

The example shows that the observed combined effect of the two cholinesterase-inhibitors is not significantly different from an expected additive response but significantly greater than expected for an independent interaction.

Example 2a. Graded response
Evaluation with ALLFIT

Construction of experimental and theoretical DRCs
As an example for a construction of theoretical DRCs by ALLFIT as well as for a quantitative response, the relaxant effect of salbutamol in the absence and presence of a fixed dose of verapamil on bovine tracheal smooth muscle is evaluated here. Values for percent relaxation are given in Table 11 (further examples in Tables 16 and 17).

From the individual values of 10 and 7 smooth muscles strips, respectively, median values ($\bar{x}$) are first determined, e.g., by STATGRAPHICS (Estimation and Testing). Then, a file of the dose-effect values is created according to the User's Guide to ALLFIT (De Lean et al. 1988), e.g., in WORD (Microsoft Word):

*** SaVe Salbutamol ***
0,0
0.1,3.5
1,29
50,70
200,73

*** plus 100 ng Verapamil ***
0,38
0.1,48
1,68
10,82
50,84
200,87

This file is saved as an unformatted ALLFIT file (text only) with the extension ".all", in this example as "c:\allfit\SaVe.all", since ALLFIT has been installed on the hard disc (c:).

Then, the program ALLFIT can be executed according to the instructions, and a graph file is created, e.g., as "SaVe". After changing to the GRAFIT program, the graph file is used for creating a graph, in this case by the command "c:\allfit\save". The program selects values for the scales, which can be changed. For the printout in Fig. 52a the following values were used: xmin = -2 (note that in the program minus is indicated by a hyphen), xmax = 4, delta x = 1; ymin = 0, ymax = 100, delta y = 10.

We can now check the curve-fitting procedure, and then look at the parameters we need to know. The parameters of the curves are indicated in the ALLFIT program by "a", "b", "c", and "d", indicating E_{max}, slope, ED_{50}, and E_{min}. The parameters obtained with curves 1 and 2 were:

Table 11. Percent relaxation of salbutamol ($\mu g/ml$) in the absence and presence of a fixed dose of verapamil (100 ng/ml) on 10 and 7 smooth muscle strips, respectively

0	0.1	1	10	50	200
0	8	41	66	70	73
0	16	52	76	84	84
0	0	3	38	54	63
0	4	34	63	80	80
0	30	29	66	78	81
0	9	41	66	70	70
0	4	29	62	66	73
0	0	14	50	64	72
0	0	8	56	76	82
0	0	3	38	54	67
$\tilde{x} = $ 0	3.5	29	62.5	70	73

plus verapamil

34	41	66	80	88	90
25	32	57	82	84	89
42	46	67	79	83	83
48	64	77	87	90	90
50	55	72	85	87	87
36	67	72	74	76	76
38	48	68	83	84	84
$\tilde{x} = $ 38	48	68	82	84	87

a1 = 72.8	a2 = 86.5
b1 = −0.99	b2 = −0.78
c1 = 1.56	c2 = 0.55
d1 = 0	d2 = 38

For the construction of the theoretical additive DRC, we need E_{max} (a1), slope (b1), and the ED_{50} (c1) of the curve of A (salbutamol) as well as E_{min} (d2) of the curve of B (salbutamol in the presence of verapamil). Hence, we look at these values, conveniently by using the NORTON COMMANDER (Symantex Corp.). These values have been indicated in Fig. 52a. According to the principle of construction of the theoretical DRCs, *and* by taking advantage of Tables 12 and 13, we can proceed as follows.

First, we supplement the dose-effect file by indicating two doses each for the minimum and the maximum of the theoretical curves, again most conveniently by using the NORTON COMMANDER command F4 (edit):

```
*** addit ***
0.001,38
0.002,38
500,73
1000,73
*** indep ***
0.001,38
0.002,38
500,83
1000,83
```

Having completed the file, we save it (command F2). The doses chosen for the minima and maxima of the theoretical curves should be clearly below the expected minimum of d2 (38%) and above E_{max} of A (salbutamol), respectively. The maximum effect expected for an independent combination has to be calculated, analogously as described for Example 1. The value in this example is 83%.

Use of Tables 12 and 13
The slope- and ED_{50}-values for the additive DRC can be derived from Tables 12 and 13, respectively. In case that substance A (salbutamol) does not reach the possible maximum of 100 %, we have to calculate a *corrected value for B* (verapamil) to read off the respective values for the slope of the additive DRC from Table 12 and the multiplication factor for the ED_{50} of the additive DRC from Table 13. The corrected value for B is the value of percentage relative to E_{max} of substance A, here:

$$B_{corr} = 38 \times 1.376 = 52.3.$$

From Table 12 we can see that the slope of the additive DRC we want to create must be close to 1 in this example, −1 in ALLFIT! From Table 13 we can see that the factor for calculating the additive ED_{50} is between 2.00 and

Table 12. Slope of additive DRC as a function of the corrected effect of B and the slope of the control curve A. E_B corr = $E_B \times 100$ / E_{max} of A

E_B corr	\multicolumn{19}{c}{S l o p e o f A}
	0.6	0.7	0.8	0.9	1.0	1.1	1.2	1.3	1.4	1.5	1.6	1.7	1.8	1.9	2.0	2.1	2.2	2.3	2.4
5	0.62	0.72	0.82	0.91	1.00	1.09	1.17	1.25	1.33	1.40	1.47	1.54	1.60	1.67	1.73	1.79	1.85	1.90	1.96
10	0.64	0.73	0.83	0.90	1.00	1.08	1.15	1.22	1.29	1.36	1.42	1.48	1.53	1.59	1.64	1.69	1.75	1.80	1.84
15	0.66	0.75	0.84	0.91	1.00	1.08	1.14	1.20	1.27	1.33	1.38	1.44	1.49	1.54	1.57	1.64	1.68	1.71	1.79
20	0.67	0.76	0.85	0.91	1.00	1.07	1.13	1.19	1.25	1.31	1.36	1.41	1.45	1.48	1.50	1.58	1.62	1.66	1.70
25	0.67	0.77	0.86	0.91	1.00	1.07	1.13	1.18	1.24	1.29	1.34	1.38	1.43	1.46	1.49	1.55	1.59	1.63	1.66
30	0.67	0.78	0.86	0.91	1.00	1.06	1.12	1.17	1.22	1.27	1.31	1.35	1.40	1.44	1.48	1.52	1.56	1.59	1.62
35	0.69	0.79	0.87	0.92	1.00	1.06	1.12	1.17	1.22	1.26	1.30	1.34	1.38	1.42	1.46	1.50	1.55	1.57	1.60
40	0.71	0.79	0.87	0.92	1.00	1.06	1.11	1.16	1.21	1.24	1.29	1.33	1.36	1.40	1.44	1.48	1.51	1.54	1.57
45	0.72	0.80	0.88	0.93	1.00	1.06	1.11	1.16	1.20	1.24	1.28	1.32	1.35	1.39	1.42	1.46	1.49	1.52	1.54
50	0.73	0.80	0.88	0.94	1.00	1.05	1.10	1.15	1.19	1.23	1.27	1.30	1.34	1.37	1.40	1.44	1.47	1.50	1.52
55	0.74	0.81	0.89	0.94	1.00	1.05	1.10	1.15	1.19	1.22	1.26	1.29	1.33	1.39	1.42	1.46	1.49	1.51	1.59
60	0.75	0.82	0.89	0.94	1.00	1.05	1.09	1.14	1.18	1.21	1.25	1.28	0.91	1.35	1.38	1.41	1.44	1.47	1.49
65	0.76	0.82	0.89	0.95	1.00	1.05	1.09	1.14	1.17	1.21	1.24	1.27	0.96	1.34	1.37	1.40	1.43	1.46	1.48
70	0.76	0.82	0.89	0.95	1.00	1.05	1.09	1.13	1.17	1.20	1.23	1.26	1.02	1.32	1.36	1.38	1.41	1.44	1.46
72.5	0.76	0.83	0.89	0.95	1.00	1.05	1.09	1.13	1.16	1.20	1.23	1.26	1.06	1.32	1.35	1.38	1.41	1.43	1.46
75	0.77	0.83	0.90	0.95	1.00	1.05	1.09	1.13	1.16	1.20	1.23	1.26	1.11	1.32	1.35	1.38	1.40	1.43	1.45
77.5	0.77	0.83	0.90	0.95	1.00	1.05	1.09	1.12	1.16	1.19	1.23	1.26	1.16	1.32	1.34	1.37	1.40	1.42	1.45
80	0.77	0.83	0.90	0.95	1.00	1.04	1.09	1.12	1.16	1.19	1.22	1.25	1.23	1.31	1.34	1.37	1.40	1.42	1.44
82.5	0.77	0.84	0.90	0.95	1.00	1.04	1.08	1.12	1.16	1.19	1.22	1.25	1.31	1.31	1.34	1.36	1.39	1.41	1.44
85	0.78	0.84	0.90	0.95	1.00	1.04	1.08	1.12	1.15	1.19	1.22	1.25	1.42	1.31	1.33	1.36	1.39	1.41	1.43
87.5	0.78	0.84	0.90	0.95	1.00	1.04	1.08	1.12	1.15	1.19	1.22	1.25	1.55	1.30	1.33	1.35	1.38	1.40	1.43
90	0.78	0.84	0.90	0.95	1.00	1.04	1.08	1.12	1.15	1.18	1.21	1.25	1.74	1.30	1.32	1.35	1.38	1.40	1.42
92.5	0.78	0.85	0.90	0.95	1.00	1.04	1.08	1.11	1.15	1.18	1.21	1.24	2.03	1.30	1.32	1.35	1.37	1.40	1.42
95	0.79	0.85	0.91	0.95	1.00	1.04	1.08	1.11	1.15	1.18	1.21	1.24	2.52	1.29	1.31	1.35	1.37	1.39	1.42

2.22 for the corrected E_B between 50 and 55 % . Hence, the ED_{50} of the additive DRC can be calculated by multiplying the ED_{50} of substance A (salbutamol), i.e., c1 with about 2.1, i.e., $1.56 \times 2.1 = 3.3$. No such calculations are necessary for the the slope and ED_{50} of the independent DRC, since it must share these values with the curve of substance A!

We can now repeat the curve-fitting procedure with the extended file "SaVe.all" and take advantage of the features of ALLFIT. The additive DRC corresponds to curve 3, the independent curve (in this example) to curve 4. We therefore indicate the *shared parameters* as follows:

a1 a3; b1 b4; c1 c4; d2 d3 d4.

Table 13. Multiplication factor for calculation of the ED_{50} of the additive DRC as a function of the corrected effect of B and the slope of the control curve A. E_B corr = $E_B \times 100 / E_{max}$ of A

E_B corr	Slope of A																		
	0.6	0.7	0.8	0.9	1.0	1.1	1.2	1.3	1.4	1.5	1.6	1.7	1.8	1.9	2.0	2.1	2.2	2.3	2.4
5	1.18	1.14	1.11	1.08	1.05	1.03	1.00	0.98	0.95	0.93	0.91	0.88	0.86	0.84	0.82	0.80	0.79	0.77	0.75
10	1.37	1.29	1.22	1.16	1.11	1.06	1.02	0.98	0.94	0.91	0.88	0.85	0.82	0.80	0.77	0.75	0.73	0.71	0.69
15	1.60	1.46	1.35	1.25	1.18	1.11	1.05	1.00	0.95	0.91	0.87	0.84	0.80	0.77	0.74	0.72	0.70	0.67	0.65
20	1.87	1.65	1.48	1.36	1.25	1.16	1.09	1.02	0.97	0.91	0.87	0.83	0.79	0.76	0.73	0.70	0.67	0.65	0.62
25	2.18	1.87	1.64	1.47	1.33	1.22	1.13	1.05	0.98	0.93	0.87	0.83	0.79	0.76	0.71	0.68	0.65	0.63	0.60
30	2.56	2.12	1.82	1.60	1.43	1.29	1.18	1.09	1.00	0.94	0.88	0.83	0.79	0.75	0.71	0.68	0.65	0.62	0.59
35	3.03	2.43	2.03	1.75	1.54	1.37	1.24	1.13	1.04	0.97	0.90	0.84	0.79	0.75	0.71	0.67	0.64	0.61	0.58
40	3.60	2.80	2.28	1.93	1.67	1.47	1.31	1.19	1.10	1.00	0.92	0.86	0.80	0.75	0.71	0.67	0.64	0.61	0.58
45	4.32	3.24	2.58	2.14	1.82	1.58	1.40	1.25	1.13	1.03	0.95	0.88	0.82	0.77	0.72	0.68	0.64	0.61	0.58
50	5.24	3.80	2.95	2.39	2.00	1.72	1.50	1.33	1.19	1.08	0.99	0.91	0.84	0.78	0.73	0.69	0.65	0.61	0.58
55	6.46	4.52	3.41	2.70	2.22	1.88	1.62	1.42	1.27	1.14	1.03	0.95	0.87	0.81	0.75	0.70	0.66	0.62	0.59
60	8.11	5.46	4.00	3.10	2.50	2.08	1.77	1.54	1.36	1.21	1.09	0.99	0.91	0.84	0.78	0.72	0.68	0.63	0.60
65	10.5	6.74	4.78	3.61	2.86	2.34	1.97	1.69	1.47	1.30	1.16	1.05	0.96	0.88	0.81	0.75	0.70	0.65	0.61
70	13.9	8.56	5.86	4.31	3.33	2.68	2.22	1.88	1.62	1.42	1.26	1.13	1.02	0.93	0.85	0.79	0.73	0.68	0.64
72.5	16.3	9.79	6.84	4.76	3.64	2.89	2.38	2.00	1.71	1.49	1.32	1.18	1.06	0.96	0.89	0.81	0.75	0.70	0.65
75	19.4	11.3	7.44	5.30	4.00	3.15	2.56	2.14	1.82	1.58	1.39	1.23	1.11	1.00	0.91	0.84	0.77	0.72	0.67
77.5	23.4	13.3	8.53	5.97	4.45	3.46	2.79	2.31	1.95	1.68	1.47	1.30	1.16	1.05	0.95	0.87	0.80	0.74	0.69
80	28.9	15.8	9.93	6.82	5.00	3.84	3.07	2.52	2.11	1.81	1.57	1.38	1.23	1.05	1.00	0.91	0.83	0.77	0.72
82.5	36.5	19.3	11.79	7.93	5.72	4.33	3.41	2.78	2.31	1.96	1.69	1.48	1.31	1.17	1.06	0.96	0.88	0.81	0.75
85	47.8	24.3	14.4	9.43	6.67	4.98	3.87	3.11	2.56	2.16	1.85	1.61	1.42	1.26	1.13	1.02	0.93	0.86	0.79
87.5	65.6	31.8	18.13	11.6	8.00	5.86	4.49	3.56	2.90	2.42	2.06	1.78	1.55	1.37	1.23	1.11	1.00	0.92	0.84
90	96.4	44.0	24.1	14.9	10.00	7.17	5.39	4.21	3.39	2.79	2.35	2.01	1.74	1.53	1.36	1.22	1.10	1.00	0.91
92.5	156.0	66.9	34.7	20.5	13.3	9.29	6.83	5.23	4.14	3.36	2.79	2.36	2.03	1.77	1.55	1.38	1.24	1.12	1.02
95	313	120	57.8	32.2	20.0	13.4	9.55	7.11	5.50	4.38	3.57	2.98	2.52	2.17	1.89	1.66	1.47	1.32	1.19

Curve 3, the additive DRC must share E_{max} with curve 1; curve 4, the independent curve, must share slope and ED_{50} with curve 1. All curves in combination, whether observed (curve 2) or calculated (curves 3 and 4) must have the same E_{min}-value.

We then indicate the *constant parameters*:

a1 a4 b1 b3 c1 c3 d2.

After pressing the ENTER key, we indicate the values themselves, i.e., 73 for a1, a3; 83 for a4; -1 for b1 as well as b3; 1.56 for c1, c4; 3.3 for c3, and 38 for d2, d3, d4. Although it is not necessary to indicate b1 and c1, this will make

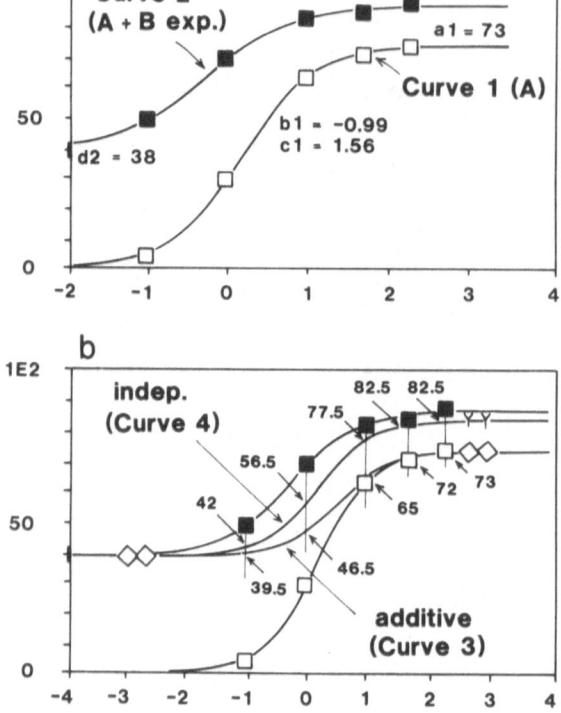

Fig. 52. Illustration of construction of theoretical DRCs with ALLFIT: Example 2a. Fitted curves of experimental effects with the parameters indicated (**a**), which are used to create the theoretical DRCs (**b**) as described in the text. The small arrows in **b**) point to the intersecting lines, i.e., to the estimated effects of the theoretical curves 3 and 4

sure that the program will not fit the DRC of substance A with the 3 curves differently to the fit of the experimental curves (in Fig. 52a).

Statistical evaluation

The extended graph file will then create a graph, illustrated in Fig. 52b. From this graph (or by other means) we then determine the values of the additive and independent DRC of interest, e.g., as illustrated in Fig. 52b. As the experimental median values ($\bar{x}$) are above the theoretical DRCs, we look at the number N of individual experimental (observed) values above them (Table 11). (Otherwise we look at the numbers below the theoretical curves, in order to have higher observed numbers for comparison with expected numbers.)

Then compare the observed values of N with the expected values. The latter is half of the total N. We can indicate the median values of the additive and the independent DRC and the observed and expected numbers, as we type them into STATGRAPHICS (Tables 14 and 15).

The calculation for additive action yields $\chi^2 = 15.78$, 4 d.f.; p = 3.3 E – 3 = 0.0033, hence a significant overadditive combination.

Table 14. Additive

Median effects (%)	40	46.5	65	72	73
Obs. numbers above	6	7	7	7	7
Exp. numbers above	3.5	3.5	3.5	3.5	3.5

Table 15. Independent

Median effects (%)	41	56.5	77.5	82.5	82.5
Obs. numbers above	5	7	6	6	6
Exp. numbers above	3.5	3.5	3.5	3.5	3.5

The statistical evaluation for independent action yielded $\chi^2 = 9.5$, 4 d.f.; $p = 0.05$. Hence, the greater than independent effect is at the border of significance.

Example 2b. Graded response
Evaluation with Sigma-Plot

Construction of experimental and theoretical DRCs
Quite recently, Hans-Peter Baer in collaboration with the author developed supplementary programs for curve fittings and mathematical transformations which can be used with Sigma-Plot (Jandel Corp.) instead of ALLFIT. Interested users of Sigma-Plot can obtain these supplementary programs for PC together with instructions from the author. The principle is described here, and the obtained graph of Example 2b is shown in Fig. 53.

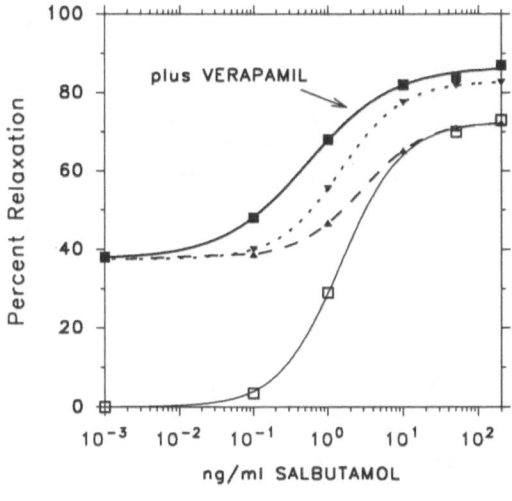

Fig. 53. Illustration of construction of theoretical DRCs with Sigma-Plot: Example 2b. Experimental and theoretical DRCs were obtained with the aid of supplementary program files referred to in the text. Laser printout of the original graph, generated by Sigma-Plot version 5.0

Table 16. Percent relaxation of 2-pentanol (μl/ml) in the absence and presence of a fixed dose of DMSO (30 μl/ml) on 12 and 8 smooth muscle strips, respectively

0	0.25	0.5	1	2	4	8	16	32
0	0	0	3	13	51	81	92	100
0	0	0	0	3	17	60	83	97
0	0	2	13	44	87	97	94	100
0	0	0	1	40	83	94	88	100
0	0	2	10	34	73	88	96	100
0	0	0	12	47	84	95	95	99
0	0	0	0	0	8	41	72	99
0	0	0	0	5	33	78	92	98
0	0	0	0	0	21	67	92	–
0	0	0	4	13	58	78	75	100
0	0	0	5	18	71	84	83	93
0	0	1	2	7	29	60	78	100
plus DMSO								
60	63	66	74	84	94	97	100	99
51	51	53	65	83	95	97	100	100
64	62	63	75	85	92	96	100	95
49	54	62	79	87	92	95	100	98
52	47	45	67	89	98	100	99	100
41	42	44	49	65	91	96	100	97
42	44	46	54	69	90	94	99	100
55	56	57	71	86	93	96	100	99

Table 17. Percent relaxation of salbutamol (μg/ml) in the absence and presence of a fixed dose of DMSO (25 μl/ml) on 11 and 8 smooth muscle strips, respectively

0	0.001	0.01	0.1	1	10
0	1	42	76	81	86
0	1	31	75	82	82
0	1	7	17	24	32
0	0	6	21	28	46
0	0	78	84	88	88
0	4	76	87	91	91
0	0	17	45	45	54
0	4	18	22	22	26
0	1	22	75	87	87
0	1	16	88	89	91
0	2	18	43	62	75
plus DMSO					
57	64	76	83	86	89
29	32	43	75	81	82
61	66	78	88	91	94
27	33	66	82	84	87
36	49	83	89	90	92
53	66	82	87	91	93
32	36	58	82	91	93
32	67	83	93	96	96

The (median) x,y-values for A alone and for A + B are put into specified columns in the DRC.SPG-file. Then, three fits with supplied .FIT-files, and two transformations with .XFM-files are performed in distinct succession. The parameters of the curve fits have to be put into specified columns. Thereafter, the graph can be viewed and worked on. The worksheet also contains the y-values for additive and independent combinations for the respective x-values of A + B, on which the statistical treatment can be performed as described above. The supplementary programs accept experimental data expressed as percent of the possible maximum response or its frequency, necessary for estimation of independent effects.

The results obtained with these fits and transforms using Sigma-Plot are virtually identical with those obtained by ALLFIT. For example, the parameters for slope and ED_{50} for the first curve (salbutamol alone) were 1.06 and 1.48 with Sigma-Plot, and 0.99 and 1.56 with ALLFIT.

Further examples for practise

Examples for exercising are given in Tables 16 and 17. Since the data values represent graded responses, median values have to be calculated first, a dose-effect file with A alone and with A in the presence of B is then required. In case ALLFIT is used, the values for the slope and the multiplication factor for the ED_{50} of the additive DRC can be derived from Tables 12 and 13, once the parameters of the curves of A and A plus B have been derived. In the example of Table 16, the slope of A = −2.48, which is outside the range of Table 12 and 13. Nevertheless, the values for 2.4 could be taken with little drop in accuracy of the created theoretical curves.

The resulting experimental and theoretical DRCs, together with the statistics, are shown in Fig. 41 d and f for Tables 16 and 17, respectively.

Prediction of combined effects – choosing the proper doses for an experiment

The magnitude of enhancement of effects of A by B may be tested before the actual experiment or may be assumed to (roughly) coincide with effects of an additive or an independent interaction. The effects for these model interactions can be estimated and the proper doses for a combination experiment of A in the presence of a fixed dose of B can be chosen (see Pöch et al. 1990b).

The calculation of the theoretical curves requires the knowledge or assumption of the DRC of A and of the effect of B. The theoretical curves for Examples 1 and 2 have been derived in this way. Instead of taking experimental effects of B, we can take any effect of B for calculation of the additive and independent DRCs. In general, it is advantageous to choose doses of B with about 30–40% effect of the maximum. At these effects of B, a separation between additive and independent DRCs will be best possible, if possible at all (see also Pöch et al. 1990a). The relationship between additive and independent DRCs has been described in Chap. 3 and illustrated in Fig. 13.

Tips and comments

Choosing the proper doses for a combination experiment has been outlined above. The examples for self examination and practising have been provided for the construction of theoretical curves for both, additive and independent combinations, irrespective that in all examples, except experiment 1, no competitive interaction at a common binding site can be assumed to occur. Hence, the *calculation of additivity* should be considered as demonstration and practise for those situations where drugs may share the same molecular site of action.

The *calculation of independent effects* is much easier and more meaningful in cases where substances act at different molecular sites – but is also of interest in cases where a competition for the same binding site is tested. In the latter case we can *quantitatively* compare non-additive combinations with independent effects. If ALLFIT is used, we obtain in a table not only the parameter estimates a–d but also the ratios of the ED_{50}s as relative potency estimates of the ED_{50}s of DRCs compared with the ED_{50} of curve 1. Thus, in experiments like those above we obtain the dose ratio (DR) of the shift of the DRCs, which we can consider as measure of potentiation and antagonism, respectively (see Chap. 3). So, when we are only interested in the shift of DRCs of A caused by B, we need not calculate independent effects since the ED_{50} of the independent DRC equals the ED_{50} of A alone.

In a site-directed *qualitative* study, it is also useful to compare experimental DRCs with curves of independent combinations. Thereby, we can check whether additive combinations could be brought about by independent mechanism, i.e., by compounds acting at different sites (see Pöch et al. 1990a).

Curve-fitting procedures yield the *slope of DRCs*. Note that the slope values obtained by ALLFIT are negative when the DRC rises and positive when the curve falls.

It is advisable to set d = 0 where DRCs of A have to start at 0%, also a = 100 can be fixed as a constant parameter where we know that 100% will be reached. Sometimes, a > 100 will be obtained in fitting. If this cannot occur, we set a = 100 either from the beginning or after interruption of the fitting procedure by pressing the control plus the "C" key, and restarting the program. If we want to know whether fitting will yield greater values than 100, we can choose 2 or 3 numbers of iterations only; the program will then stop after the number of chosen iterations, and we can choose additional iterations thereafter. Regardless of the number of iterations chosen, the program will end when the best fit has been obtained: this may sometimes happen after a few iterations.

References

Adashi EY, Resnick CE, Svoboda ME, Van Wyk JJ (1985) Somatomedin-C synergizes with follicle-stimulating hormone in the acquisition of progestin biosynthetic capacity by cultured rat granulosa cells. Endocrinology 116: 2135–2142

Agarwal AK, Mehendale HM (1984) Excessive hepatic accumulation of intracellular Ca^{2+} in chlordecone potentiated CCl4 toxicity. Toxicology 30: 17–24

Akobundu IO, Sweet RD, Duke WB (1975) A method of evaluating herbicide combinations and determining herbicide synergism. Weed Sci 23: 20–25

Altenburger R, Bödecker W, Faust M, Grimme LH (1990) Evaluation of the isobologram method for the assessment of mixtures of chemicals. Combination effects studies with pesticides in algal biotests. Ecotox Environ Safety 20: 98–114

– – – – (1992) Analysis of combination effects in aquatic toxicology. In: Corn M (ed) Handbook of hazardous materials. Academic Press, San Diego (in press)

Amyes S (1982) Bactericidal activity of trimethoprim alone and in combination with sulfamethoxazole on susceptible and resistant *Escherichia coli* K-12. Antimicrob Agents Chemother 21: 288–293

Ankenbauer W, Strähle U, Schütz G (1988) Synergistic action of glucocorticoid and estradiol responsive elements. Proc Natl Acad Sci USA 85: 7526–7530

Arcos JC, Woo Y-T, Lai DY (1988) Database on binary combination effects of chemical carcinogens. J Environ Sci Health C6: 1–150

Ariëns EJ (1979) Receptors: from fiction to fact. Trends Pharmacol Sci 1: 11–15

– Van Rossum JM, Simonis AM (1956a) A theoretical basis of molecular pharmacology. Part I: Interactions of one or two compounds with one receptor system. Arzneimittelforschung 6: 282–293

– – – (1956b) A theoretical basis of molecular pharmacology. Part III: Interactions of one or two compounds with two independent receptor systems. Arzneimittelforschung 6: 737–746

– Simonis AM, Van Rossum JM (1964a) Drug-receptor interaction: interaction of one or more drugs with one receptor system. In: Ariëns EJ (ed) Molecular pharmacology. The mode of action of biologically active compounds. Academic Press, New York, pp 119–286

– – – (1964b) The relation between stimulus and effect. In: Ariëns EJ (ed) Molecular pharmacology. The mode of action of biologically active compounds. Academic Press, New York, pp 394–466

– Beld AJ, Rodrigues de Miranda, Simonis AM (1979) The pharmacon-receptor-effector concept. In: O'Brien (ed) The receptors. A comprehensive treatise. Plenum, New York, pp 33–91

Arunlakshana O, Schild HO (1959). Some quantitative uses of drug antagonists. Br J Pharmacol 14: 48–58

Ashford JR, Smith CS (1964) General models for quantal response to the joint action of a mixture of drugs. Biometrika 51: 413–428

Ben-Shlomo I, Abd-El-Khalim H, Ezry J, Zohar S, Tverskoy M (1990) Midazolam acts synergistically with fentanyl for induction of anaesthesia. Br J Anaesth 64: 45–47

Berenbaum MC (1977) Synergy, additivism and antagonism in immunosuppression. A critical review. Clin Exp Immunol 28: 1–18

Berenbaum MC (1978) A method for testing for synergy with any number of agents. J Infect
    Dis 137: 122–130
–   (1981) Criteria for analysing interactions between biologically active agents. Adv Cancer
    Res 35: 269–335
–   (1980) Correlations between methods for measurement of synergy. J Infect Dis 142: 476–
    479
–   (1985) The expected effect of a combination of agents: the general solution. J Theor Biol
    114: 413–431
–   (1988) Isobolographic, algebraic, and search methods in the analysis of multiagent
    synergy. J Am Coll Toxicol 7: 927–938
–   (1989) What is synergy? Pharmacol Rev 41:93–141
–   Yu VL, Feegie TP (1982) Synergy with double and triple antibiotic combinations com-
    pared. J Antimicrob Chemother 12: 555–563
Berger MR, Schmähl D, Edler L (1990) Implications of the carcinogenic hazard of low doses
    of three hepatocarcinogenic N-nitrosamines. Jpn J Cancer Res 81: 598–606
Birgersson B, Sterner O, Zimerson E (1988) Chemie und Gesundheit. Verlag Chemie,
    Weinberg, p 157
Black JW, Leff P (1983) Operational models of pharmacological agonism. Proc R Soc Lond
    [Biol] 220: 141–162
–   – Shankley NP, Wood J (1985) An operationel model of pharmacological agonism: the
    effect of E/[A] curve shape on agonist dissociation constant estimation. Br J Pharmacol
    84: 561–571
Black ML (1963) Sequential blockage as a theoretical basis for drug synergism. J Med Chem
    6: 145–153
Bliss CI (1939) The toxicity of poisons applied jointly. Ann Appl Biol 26: 585–615
Bödeker W, Altenburger R, Faust M, Grimme LH (1990) Methods for the assessment of
    mixtures of plant protection substances (pesticides): mathematical analysis of combi-
    nation effects in phtyopharmacology and ecotoxicology. Nachrichtenbl Deutsch
    Pflanzenschutzd 42: 70–78
Bosma A, Brouwer A, Seifert WF, Knook DL (1988) Synergism between ethanol and carbon
    tetrachloride in the generation of liver fibrosis. J Pathol 156: 15–21
Bourne HR, Roberts JM (1987) Drug receptors and pharmacodynamics. In: Katzung BG (ed)
    Basic and clinical pharmacology. Appleton & Lange, Norwalk, CT, pp 9–22
Bowman WC, Bowman A, Bowman A (1986) Dictionary of pharmacology. Blackwell, Oxford
Brunden MN, Vidmar TJ, McKean JW (1988) Drug interaction and lethality analysis. CRC
    Press, Boca Raton
Brunet BL, Reiffenstein RJ, Williams T, Wong L (1986) Toxicity of phencyclidine and
    ethanol in combination. Alcohol Drug Res 6: 341–349
Bürgi E (1938) Die Arzneikombinationen. Springer, Berlin
Buttermann G, Theisinger W, Weidenbach A, Hartung R, Welzel D, Pabst HW (1977) Quan-
    titative Bewertung der postoperativen Thromboembolieprophylaxe. Med Klin 72: 1624–
    38
Calamari D, Alabaster JS (1980) An approach to theoretical models in evaluating the effects
    of mixtures of toxicants in the aquatic environment. Chemosphere 9: 533–538
Caplan RA, Su JY (1986) Interaction of halothane and verapamil in isolated papillary muscle.
    Anesth Analg 65: 463–468
Carter WH Jr (1985) Response surface methodology and the design of clinical trials for the
    evaluation of cancer chemotherapy. Cancer Treatm Rep 69: 1049–51
–   Carchman RA (1988) Mathematical and biostatistical methods for designing and
    analyzing complex chemical interactions. Fund Appl Toxicol 10: 590–595
–   Gennings C, Staniswalis JG, Campbell ED, White KL Jr (1988) A statistical approach to the
    construction and analysis of isobolograms. J Am Coll Toxicol 7: 963–973
Chadwick RW, Scotti TM, Copeland MF, Mole ML, Froehlich R, Cooke N, McElroy WK
    (1984) Antagonism of chlorobenzene-induced hepatoxicity by lindane. Pest Biochem
    Physiol 21: 148–161

Chou J, Chou T-C (1987) Dose-effect analysis with microcomputers. Elsevier-Biosoft, Cambridge

Chou T-C, Rideout DC (1991) Synergism and antagonism in chemotherapy. Academic Press, San Diego

- Talalay P (1977) A simple generalized equation for the analysis of multiple inhibitions of Michaelis-Menten kinetic systems. J Biol Chem 252: 6438–42

- - (1981) Generalized equations for the analysis of inhibitions of Michaelis–Menten and higher-order kinetic systems with two or more mutually exclusive and nonexclusive inhibitors. Eur J Biochem 115: 207–216

- - (1983) Analysis of combined drug effects: a new look at a very old problem. Trends Pharmacol Sci 4: 450–454

- - (1984) Quantitative analysis of dose-effect relationships: the combined effects of multiple drugs or enzyme inhibitors. Adv Enzyme Regul 22: 27–55

Christensen ER, Chen Ch-Y (1985) A general noninteractive multiple toxicity model including probit, logit, and Weibull transformations. Biometrics 41, 711–725

Cleland WW (1970) Steady state kinetics. In: Boyer PD (ed) The enzymes. Kinetics and mechanism, vol II. Academic Press, New York, pp 33–35

Cohen JA, Warringa GPJ, Bovens BR (1951) Protection of true cholinesterase against diisopropyl fluorophosphonate by butyrylcholine. Biochim Biophys Acta 6: 469–476

Cohen P (1980) Recently discovered systems of enzyme regulation by reversible phosporylation. Elsevier, Amsterdam

Colby SR (1967) Calculating synergistic and antagonistic responses of herbicide combinations. Weeds 15: 20–22

Cole DJ, Kalichman MW, Shapiro HM (1989) The nonlinear contribution of nitrous oxide at sub-MAC concentrations to enflurane MAC in rats. Anesth Analg 68: 556–562

- - - Drummond JC (1990) The non-linear potency of sub-MAC concentrations of nitrous oxide in decreasing the anesthetic requirement of enflurane, halothane, and isoflurane in rats. Anesthesiology 73: 93–99

Creveling CR Van der Schoot JB, Udenfriend S (1962) Phenylethylamine isosteres as inhibitors of dopamine ß-oxidase. Biochem Biophys Res Commun 8: 215–219

Dal Monte PR, D'Imperio M, Ferri M, Fratucello, Soldato P del (1985) A combination of pirenzepine and cimetidine: a new approach to treatment of duodenal ulcer. Hepatogastroenterol 32: 126–128

Dawson DA (1991) Additive incidence of developmental malformation for xenopus embryos exposed to a mixture of ten aliphatic carboxylic acids. Teratology 44: 531–546

- Wilke TS (1991a) Initial evaluation of developmental malformationas as an end point in mixture toxicity hazard assessment for aquatic vertebrates. Ecotox Environ Safety 21: 215–226

- - (1991b) Joint toxic action of binary mixtures of osteolathyrogens at malformation-inducing concentrations for Xenopus embryos. J Appl Toxicol 11: 415–421

- - (1991c) Evaluation of the frog embryo teratogenesis assay: Xenopus (FETAX) as a model system for mixture toxicity hazard assessment. Environ Toxicol Chem 10: 941–948

Delfraissy J-F, Wallon C, Galanaud P (1988) Interferon-alpha can synergize with interleukin 2 for human in vitro antibody response. Eur J Immunol 18: 1379–1384

De Chaffoy de Courcelles D, Roevens P, De Clerck F (1988) The synergistic effect of serotonin and epinephrine at the level of signal transduction. J Cardiovasc Pharmacol 11: S107–S110

De Divitiis O, Petitto M, Di Somma S, Galderisi M, Villari B, Santomauro M, Fazio S (1984) Nitrendipine and atenolol: comparison and combination in the treatment of arterial hypertension. Arzneimittelforschung 34: 727–729

De Lean A, Rodbard D (1979) Kinetics of cooperative binding. In: O'Brien (ed) The receptors. A comprehensive treatise. Plenum, New York, pp 143–192

- Munson PJ, Rodbard D (1978) Simultaneous analysis of families of sigmoidal curves: application to bioassay, radioligand assay, and physiological dose-response curves. Am J Physiol 235: E97–E102

De Lean A, Munson PJ, Guardabasso V, Rodbard D (1988) A user's guide to ALLFIT. Simultaneous fitting of families of sigmoidal dose response curves using the four-parameter logistic function. Laboratory Theoretical and Physical Biology, National Institute of Health, Bethesda, MD

Dietz K, Mallach HJ, Schenzle D, Schmidt V, Unkelbach HD, Wolf T (1984) Untersuchungen zur Prüfung der Wechselwirkung zwischen Alkohol und einem neuen 1,4-Benzodiazepam (Metaclazepam). 4. Mitt.: Anpassung einer Dosis-Wirkungs-Fläche für additive Komponenten. Blutalkohol 21: 14–30

DiFazio CA, Brown RE, Ball CG, Heckel CG, Kennedy SS (1972) Additive effects of anesthetics and theories of anesthesia. Anesthesiology 36: 57–63

Dilger DP, Firestone LL (1990) More models described for molecular "target" of anesthetics and alcohols. Trends Pharmacol Sci 11: 431–432

Draskoczy PR, Trendelenburg U (1968) The uptake of l- and d-noradrenaline by the isolated perfused rabbit heart in relation to the stereospecificity of the sensitizing action of cocaine. J Pharmacol Exp Ther 159: 66–73

Drewinko B, Loo TL, Brown B, Gottlieb JA, Freireich EJ (1976) Combination chemotherapy in vitro with adriamycin. Observations of additive, antagonistic, and synergistic effects when used in two-drug combinations on cultured human lymphoma cells. Cancer Biochem Biophys 1: 187–195

Ehlert FJ (1986) 'Inverse agonists', cooperativity and drug action at benzodiazepine receptors. Trends Pharmacol Sci 7: 28–32

–   (1988) Estimation of the affinities of allosteric ligands using radioligand binding and pharmacological null methods. Mol Pharmacol 33: 187–194

Elashoff RM, Fears TR, Schneiderman MA (1987) Statistical analysis of a carcinogen mixture experiment. I. Liver carcinogens. J Natl Cancer Inst 79: 509–526

Elion GB, Singer S, Hitchings GH (1954) Antagonists of nucleic acid derivatives. VIII. Synergism in combinations of biochemically related antimetabolites. J Biol Chem 208: 477–488

Finney DJ (1942) The analysis of toxicity tests on mixtures of poisons. Ann Appl Biol 29: 82–94

Fredholm BB, Brodin K, Strandberg K (1979) On the mechanism of relaxation of tracheal muscle by theophylline and other cyclic nucleotide phosphodiesterase inhibitors. Acta Pharmacol Toxicol 45: 336–344

Frei W (1913) Versuche über Kombinationen von Desinfektionsmitteln. Z Hyg Infektionskrank 75: 433–496

Freis ED (1983) Efficacy of nadolol alone and combined with bendroflumethiazide and hydralazine for systemic hypertension. Am J Cardiol 52: 1230–37

Gero A (1971) Intimate study of drug action III: Mechanisms of molecular drug action. In: DiPalma JR (ed) Drill's pharmacology in medicine, 4th edn. McGraw-Hill, New York, pp 67–98

Gessner PK (1988) A straightforward method for the study of drug interactions: an isobolographic analysis primer. J Am Coll Toxicol 7: 987–1012

Gibbins IL (1989) Co-existence and co-function. In: Holmgren S (ed) The comparative physiology of regulatory peptides. Chapman and Hall, London, pp 308–343

Goldstein RS, Hewitt WR, Hook JB (eds) (1990) Toxic interactions. Academic Press, San Diego

Gonzales RA, Hoffman PL (1991) Receptor-gated ion channels may be selective CNS targets for ethanol. Trends Pharmacol Sci 12: 1–3

Greaves MF (1976) Cell surface receptors: a biological perspective. In Cuatrecasas P, Greaves MF (eds) Receptors and recognition, series A, vol 1. Chapman and Hall, London, pp 3–32

Greco WR (1987) The assessment of synergism, antagonism and additivity: a unified optimal approach. Proc Am Assoc Cancer Res 28: 1687

–   Lawrence DD (1988) Assessment of the degree of drug interaction where the response variable is discrete. Proc Biophys Sect Am Statist Assoc 1988: 226–231

- Park HS, Rustum YM (1990) Application of a new approach for the quantitation of drug synergism to the combination of cis-diamminedichloroplatinum and 1-β-D-arabinofuranosylcytosine. Cancer Res 50: 5318–27

Green AL, Lord J, Marshall IG (1978) The relationship between cholinesterase inhibition in the chick biventer cervicis muscle and its sensitivity to exogenous acetylcholine. J Pharm Pharmacol 30: 426–231

Green JM (1989) Herbicide antagonism at the whole plant level. Weed Technol 3: 217–226

- Bailey SP (1987) Herbicide interactions with herbicides and other agricultural chemicals. In: McWhorter CG, Gebhardt MR (eds) Methods of applying herbicides. Weed Science Society, Champaign, IL, pp 37–61

Gressel J (1990) Synergizing herbicides. Rev Weed Sci 5: 49–82

Guy HR, Hucho F (1987) The ion channel of the nicotinic acetylcholine receptor. Trends Neurosci 10: 318–321

Hall MJ, Middleton RF, Westmacott D (1983) The fractional inhibitory concentration (FIC) index as a measure of synergy. J Antimicrob Chemother 11: 427–433

Hanna CJ, Roth SH (1979) Combination bronchodilators: antagonism of airway smooth muscle contractions in vitro. Agents Actions 9: 18–23

Harrap KR, Jackson RC (1975) Enzyme kinetics and combination chemotherapy: an appraisal of current concepts. Adv Enzyme Regul 13: 77–96

Harris CM, Springfield AC, Ranschaert ER, Lal H (1986) Lethal drug interaction: isoniazid and methylxanthines. Drug Dev Res 9: 299–304

Hayashi R, Suzuki S, Watanabe H, Kenmochi T, Fukuoka T, Niiya S, Amemiya H (1990) Synergistic effect of cyclosporine and mizoribine on graft survival in canine organ transplantation. Tranplant Proc 22: 1676–1678

Heagle AS, Johnston JW (1979) Variable responses of soybeans to mixtures of ozone and sulfur dioxide. J Air Pollut Control Assoc 29: 729–732

Hewlett PS, Plackett RL (1950) Statistical aspects of the independent joint action of poisons, particularly insecticides. II. Examination of data for agreement with the hypothesis. Ann Appl Biol 37: 527–552

- - (1956) The relation between quantal and graded responses to drugs. Biometrics 12: 72–78

- - (1979) The interpretation of quantal responses in biology. Edward Arnold, London

Hinson JA, Roberts DW (1992) Role of covalent and noncovalent interactions in cell toxicity: effects on proteins. Annu Rev Pharmacol Toxicol 32: 471–510

Hirschelmann R, Pöch G, Rafler I, Rickinger O, Giessler J (1991) Steroid-saving potency of nonsteroidal antiinflammatory agents – a reevaluation with the new agent CGP 28238 in rat inflammatory models. In: Pernham J, Bray MA, Van den Berg WB (eds) Drugs in inflammation. Agents Actions [Suppl] 32: 51–57

Hoel DG (1987) Statistical aspects of chemical mixtures. In: Vouk VB, Butler GC, Upton AC, Parke DV, Asher SC (eds) Methods for assessing the effects of mixtures of chemicals. Wiley, Chichester, pp 369–377

Hofmann J, Ueberall F, Posch L, Maly K, Herrmann BJ, Grunicke H (1989) Synergistic enhancement of the antiproliferative activity of cis-diamminedichloroplatinum (II), by the ether lipid analogue BM41440, an inhibitor of protein kinase C. Lipids 24: 312–317

Holzmann S, Kukovetz WR, Braida C, Pöch G (1992) Pharmacological interaction experiments differentiate between glibenclamide-sensitive K^+-changes and cyclic GMP as components of vasodilation by nicorandil. Eur J Pharmacol 215: 1–7

Hu WY, Reiffenstein RJ, Wong L (1986) Interaction between flurazepam and ethanol. Alcohol Drug Res 7: 107–117

Joly V, Bergeron Y, Bergeron MG, Carbon C (1991) Endotoxin-tobramycin additive toxicity on renal proximal tubular cells in culture. Antimicrob Agents Chemother 35: 351–357

Jungermann K, Möhler H (1980) Biochemie. Springer, Berlin Heidelberg New York

Kaltenbach M, Becker HJ, Graef V, Hunscha H (1970) Zur Therapie der Angina pectoris mit Beta-Rezeptorenblockern. Med Klin 65: 494–500

Katzung BG (1987) Basic and clinical pharmacology, 3rd edn. Appleton & Lange, Norwalk, CT

Kaumann AJ, Wittmann R, Birnbaumer L, Hoppe BH (1977) Activation of myocardial β-adrenoceptors by the nitrogen-free low affinity ligand 3′,4′-dihydroxy-α-methyl-propiophenone (U-0521). Naunyn Schmiedebergs Arch Pharmacol 296: 217–228

Kayanakis JG, Baulac L (1987) Comparative study of once-daily administration of captopril 50 mg, hydrochlorothiazide 25 mg and their combination in mild to moderate hypertension. Br J Clin Pharmacol 23: 89S–92S

Kenakin TP (1987) Pharmacologic analysis of drug-receptor interaction. Raven, New York

Kim B, Warnaka P, Konrad C (1990) Tamoxifen potentiates in vivo antitumor activity of interleukin-2. Surgery 108: 139–145

Kim SG, Mercando AD, Fisher JD (1987) Combination of tocainidine and quinidine for better tolerance and additive effects in patients with coronary artery disease. J Am Coll Cardiol 9: 1369–1374

King LJ, Parke DV (1987) Induction, inhibition and receptor interactions. In: Vouk VB, Butler GC, Upton AC, Parke DV, Asher SC (eds) Methods for assessing the effects of mixtures of chemicals. Wiley, Chichester, pp 423–446

King MM, Carlson GM (1981) Synergistic effect of Ca^{2+} and Mg^{2+} in promoting an activity of phosphorylase kinase that is insensitive to ehylene glycol bis (β-aminoethylether)-N,N′-tetraacetic acid. Arch Biochem Biophys 209: 517–523

Kissin I, Brown PT, Bradley Jr EL (1990) Morphine and fentanyl anesthetic interactions with diazepam: relative antagonism in rats. Anesth Analg 71: 236–241

Klaassen CD, Eaton DL (1991) Principles of toxicology. In: Amdur MO, Doull J, Klaassen CD (eds) Casarett and Doull's toxicology. The basic science of poisons, 4th edn. Pergamon, New York, pp 12–49

Klastersky J, Staquet MJ (eds) (1982) Combination antibiotic therapy in the compromised host. Raven, New York (Monograph series of the European Organization for Research on Treatment of Cancer, vol 9)

Könemann H (1981) Quantitative structure-activity relationships in fish toxicity studies. Toxicology 19: 209–221

Kremer AB, Egan RM, Sable HZ (1980) The active site of transketolase. Two arginine residues are essential for activity. J Biol Chem 255: 2405–10

Kukovetz WR, Pöch G, Wurm A, Holzmann S, Paietta E (1976) Effect of phosphodiesterase-inhibition on smooth muscle tone. In: Betz E (ed) Ionic actions on vasular smooth muscle. Springer, Berlin Heidelberg New York, pp 124–131

– Holzmann S, Braida C, Pöch G (1991) Dual mechanism of the relaxing effect of nicorandil by stimulation of cylic GMP formation and by hyperpolarization. J Cardiovasc Pharmacol 17: 627–633

Kundi M (1987) Evaluation models for combined effects. In: Okada A, Manninnen O (eds) Recent advances in researches on the combined effects of environmental factors. Kyoei, Kanazawa, pp 457–471

Künstler K, Klein RG (1989) Mathematical approaches in combination effects. In: Schmähl D (ed) Combination effects in chemical carcinogenesis. Verlag Chemie, Weinheim, pp 31–43

Leff P (1987) An analysis of amplifying and potentiating interactions between agonists. J Pharmacol Exp Ther 243: 1035–42

Leon AS, Hunninghake DB (1983) A multiclinic double-blind comparison of timolol and hydrochlorothiazide alone and in combination in the treatment of essential hypertension. J Clin Pharmacol 23: 5–15

Levine RR (1983) Pharmacology. Drug actions and reactions, 3rd edn. Little, Brown and Company, Boston

Liew FY, Li Y, Millott S (1990) Tumor necrosis factor-α synergizes with IFN-γ in mediating killing of Leishmania major through the induction of nitric oxide. J Immunol 145: 4306–4310

Limbird LE (1986) Cell surface receptors: a short course on theory and methods. Martinus Nijhoff, Boston

Loewe S (1953) The problem of synergism and antagonism of combined drugs. Arzneimittel-forschung 3: 285–290

– Muischnek H (1926) Über Kombinationswirkungen. I. Mitteilung: Hilfsmittel der Fragestellung. Naunyn Schmiedebergs Arch Pharmakol 114: 313–326

Londong W, Londong V, Prechtl R, Weber T, Von Werder K (1980) Interactions of cimetidine and pirenzepine on peptone-stimulated gastric acid secretion in man. Scand J Gastroent 15 (S 66) 103–112

– – Ruthe C, Weizert P (1981) Complete inhibition of food-stimulated gastric acid secretion by combined application of pirenzepine and ranitidine. Gut 22: 542–548

– Hasford J, Sander R, Sommerlatte T, Überla K, Ultsch B, Weinzierl M (1982) Prevention of recurrent bleeding from gastroduodenal ulcers by combined application of cimetidine and pirenzepine: a double-blind, randomized and multicentre trial. In: Dotevall G (ed) Advances in gastroenterology with the selective antimuscarinic compound pirenzepine. Excerpta Medica, Amsterdam, pp 152–153

Mackay D (1981) An analysis of functional antagonism and synergism. Br J Pharmacol 73: 127–134

Meddings JB, Scott RB, Fick GH (1989) Analysis and comparison of sigmoidal curves: application to dose-response data. Am J Physiol 257: G 982–989

Mehendale HM (1984) Potentiation of halomethane hepatotoxicity: Chlordecone and carbon tetrachloride. Fund Appl Toxicol 4: 295–308

Mikulski SM, Viera A, Ardelt W, Menduke H, Shogen K (1990) Tamoxifen and trifluoroperazine (stelazine) potentiate cytostatic/cytotoxic effects of P-30 protein, a novel protein possessing anti-tumour activity. Cell Tissue Kinet 23: 237–246

Motulsky HJ, Ransnas LA (1987) Fitting curves to data using nonlinear regression: a practical and nonmathematical review. FASEB J 1: 365–374

Müller U, Von Felten A (1986) Synergistische Hemmung der Thrombozytenaktivierung durch paarweise Zugabe von Medikamenten. Schweiz Med Wochenschr 116: 1495–1497

Nosál'ová V, Babul'ová A, Benes L (1990) Efficacy of combined ranitidine and pentacaine treatment in experimentally induced gastric damage in rats. Agents Actions 30: 185–187

Ohashi M (1976) Studies on the mode of antagonism between adrenergic β-mimetics and β-blocking agents (III). Functional antagonism between β-mimetics and spasmogens. Jpn J Pharmacol 26: 315–323

Pancheva S (1991) Potentiating effect of ribavirin on the antiherpes activity of acyclovir. Antiviral Res 16: 151–161

Paul E, Pfeffer M, Bödeker R-H (1988) Effect of terfenadine and ranitidine on histamine and suxamethonium wheals. Eur J Pharmacol 34: 591–594

Peper K, Bradley RJ, Dreyer F (1982) The acetylcholine receptor at the neuromuscular junction. Physiol Rev 62: 1272–1340

Perkins JP, Johnson GL, Harden TK (1978) Drug-induced modification of the responsiveness of adenylate cyclase to hormones. Adv Cycl Nucleot Res 9: 19–32

Pessayre D, Cobert B, Descatoire V, Degott C, Babany G, Funck-Brentano C, Delaforge M, Dominique L (1982) Hepatoxicity of trichloroethylene-carbon tetrachloride mixtures in rats. Gastroenterology 83: 761–772

Pircio AW, Buyniski JP, Roebel LE (1978) Pharmacological effects of a combination of butorphanol and acetaminophen. Arch Int Pharmacodyn 235: 116–123

Plaa GL, Hewitt WR, Du Souich P, Caillé G, Lock S (1982) Isopropanol and acetone potentiation of carbon tetrachloride-induced hepatotoxicity: single versus repetitive pretreatments in rats. J Tox Environ Health 9: 235–250

Plackett RL, Hewlett PS (1952) Quantal responses to mixtures of poisons. J R Statist Soc B 14: 141–163

Plummer JL, Short TG (1990) Statistical modelling of the effects of drug combinations. J Pharmacol Meth 23: 297–309

Pöch G (1983) Synergismus bei physiologischen, pathologischen und therapeutischen Wirkungen. Funkt Biol Med 2: 102–110

Pöch G (1991a) Direct analysis of dose-response curves with respect to additive and independent interaction. In: Fechter LD (ed) Proc. of the 4th International Conference on the Combined Effects of Environmental Factors 1990. John Hopkins University, Baltimore, pp 1–4

– (1991b) Evaluation of combined effects with respect to independent action. Arch Compl Environ Studies 3: 65–74

– Brunner F (1984) The underlying bases of pharmacological results in agreement or disagreement with the law of initial value. Gen Pharmacol 15: 229–232

– Holzmann S (1980/1981) Quantitative estimation of overadditive and underadditive drug effects by means of theoretical additive dose response curves. J Pharmacol Methods 4: 179–188; Erratum 5: 183

– Juan H (1990) Wirkungen von Pharmaka, 2nd edn. Thieme, Stuttgart

– Londong W (1985) Simple approach to assess potentiated drug combinations in clinical trials: studies with pirenzepine plus H2-receptor antagonists. Int J Clin Pharmacol Ther Toxicol 23: 283–287

– Reiffenstein RJ (1992) Inappropriate and wrong conclusions derived from isobolograms. Arch Compl Environ Studies (in press)

– Zimmermann I (1988) Simple pA2 estimation of partial agonist: comparison with the Kaumann-Blinks method. J Pharmacol Methods 19: 47–56

– Dittrich P, Holzmann S (1990a) Evaluation of combined effects in dose-response studies by statistical comparison with additive and independent interactions. J Pharmacol Methods 24: 311–325

– – Reiffenstein RJ, Lenk W, Schuster A (1990b) Evaluation of experimental combined toxicity by use of dose-frequency curves: comparison with theoretical additivity as well as independence. Can J Physiol Pharmacol 68: 1338–1345

– Reiffenstein RJ, Unkelbach H-D (1990c) Application of the isobologram technique for the analysis of combined effects with respect to additivity as well as independence. Can J Physiol Pharmacol 68: 682–688

– Brunner F, Kühberger E (1992) Construction of antagonist dose-response curves for estimation of pA2-values by Schild-plot analysis and detection of allosteric interactions. Br J Pharmacol 106: 710–716

Prichard MN, Shipman C Jr (1990) A three-dimensional model to analyze drug-drug interactions. Antiviral Res 14: 181–206

Raftery MA, Conti-Tronconi BM, Dunn SMJ, Crawford RD, Middlemas D (1984) The nicotinic receptor: its structure, multiple binding sites, and cation transport properties. Fund Appl Toxicol 4: S34–S51

Rang HP (1960) Unspecific drug action. The effects of a homologous series of primary alcohols. Br J Pharmacol 15: 185-200

Reif AE (1985) Synergism in carcinogenesis. J Natl Cancer Inst 73: 25–39

– Colton T (1984) Determination of carcinogenic synergism form the latent period. Carcinogenesis 5: 837–840

Rideout DC, Chou T-C (1991) Synergism, antagonism, and potentiation in chemotherapy: an overview. In: Chou T-C, Rideout DC (eds) Synergism and antagonism in chemotherapy. Academic Press, San Diego, pp 3–60

Roebruck P, Unkelbach HD, Vollmar J, Köpke W, Ferner U, Fink H, Junge K, Nowak H, Widdra W (1990) Biometrische Aspekte zur Bewertung fixer Arzneimittelkombinationen. Arzneimittelforschung 40 (II): 725–729

Rollo IM (1955) The mode of action of sulphonamides, proguanil and pyrimethamine on plasmodium gallinaceum. Br J Pharmacol 10: 208–214

Rothman KJ (1974) Synergy and antagonism in cause-effect relationships. Am J Epidemiol 99: 385–388

Rothstein JL, Schreiber H (1988) Synergy between tumor necrosis factor and bacterial products causes hemorrhagic necrosis and lethal shock in normal mice. Proc Natl Acad Sci USA 85: 607–611

Rubin E, Miller KW, Roth SH (eds) (1991) Molecular and cellular mechanisms of alcohol and anesthetics. Ann NY Acad Sci 625

Ruffolo RR Jr (1982) Important concepts of receptor theory. J Auton Pharmacol 2: 277–295

Schmähl D (ed) (1988) Combination effects in chemical carcinogenesis. Verlag Chemie, Weinheim

Schultze-Werninghaus G (1981) Dosis-Wirkungs-Untersuchungen zur Frage der additiven Wirkung eines β2-Sympathikomimetikums und eines Anticholinergikums bei allergischem Asthma bronchiale. Atemweg Lungenkrank 7: 57–65

Schwarzl I (1992) Qualitative und quantitative Analyse von organischen Lösungsmitteln im Hinblick auf "unspezifische" Wirkungen am isolierten Rindertrachealmuskel. Diplomarbeit Universität Graz

Seamon KB, Daly JW (1981) Forskolin: a unique diterpene activator of cyclic AMP-generating systems. J Cycl Nucleot Res 7: 201–224

Shacter E, Chock PB, Stadtman ER (1984) Regulation through phosphorylation/dephosphorylation cascade systems. J Biol Chem 259: 12252–12259

Shertz RD, Kender WJ, Musselman RC (1980) Foliar response and growth of apple trees following exposure to ozone and sulfur dioxide. J Am Soc Hort Sci 105: 594–598

Smyth HF Jr, Weil CS, West JS, Carpenter CP (1969) An exploration of joint toxic action: twenty-seven industrial chemicals intubated in rats in all possible pairs. Tox Appl Pharmacol 14: 340–347

Sprague JB (1979) Measuremnent of pollutant toxicity to fish. II. Utilizing and applying bioassay results. Water Res 4: 3–32

Stacey NH (1987) Assessment of the toxicity of chemical mixtures with isolated rat hepatocytes: cadmium and chloroform. Fund Appl Toxicol 9: 616–622

Stara JF, Bruins R, Dourson ML, Erdreich LS, Hertzberg RC, Durkin PR, Pepelko WE (1987) Risk assessment is a developing science: approaches to improve evaluation of single chemicals and chemical mixtures. In: Vouk VB, Butler GC, Upton AC, Parke DV, Asher SC (eds) Methods for assessing the effects of mixtures of chemicals. Wiley, Chichester, pp 719–743

Steel GG (1979) Terminology in the description of drug-radiation interactions. Int J Radiol Oncol Biol Physiol 5: 1145–50

– Peckham MJ (1979) Exploitable mechanisms in combined radiotherapy-chemotherapy: The concept of additivity. Int J Radiol Oncol Biol Physiol 5: 85–91

Streibig JC (1986) Joint action of some root-absorbed herbicides in oats (Avena sativa L.). Weed Res 26: 207–214

Stroobandt R, Kesteloot H (1986) Effect of sotalol, aprindine and their combination on maximum upstroke velocity of action potential in guinea-pig papillary muscle. Eur J Pharmacol 131: 249–256

– Holvoet G, Verbeke N, Kesteloot H (1987) Effects of intravenous sotalol, aprindine and the combination of sotalol and aprindine on chronic high frequency ventricular arrhythmias in man. Eur Heart J 8: 372–377

Sühnel J (1990) Evaluation of synergism or antagonism for the combined action of antiviral agents. Antiviral Res 13: 23–40

– (1992) Zero interaction response surfaces, interaction functions and difference response surfaces for combinations of biologically active agents. Arzneimittelforschung 42: 1251–1258

Syracuse KC, Greco WR (1986) Comparison between the method of Chou and Talalay and a method for the assessment of the combined effects of drugs: a Monte-Carlo simulation study. Proc Biophys Sect Am Statist Assoc 1986: 127–132

Tallarida RJ, Raffa RB, McGonigle P (1988) Principles in general pharmacology. Springer, New York Berlin Heidelberg

– Porreca F, Cowan A (1989) Statistical analysis of drug-drug and site-site interactions with isobolograms. Life Sci 45: 947–961

Tammes P (1964) Isoboles, a graphic representation of synergism in pesticides Neth J Plant Pathol 7: 73–80

Taylor P (1990) Anticholinesterase agents. In: Gilman AG, Rall TW, Nies AS, Taylor P (eds) Goodman and Gilman's The pharmacological basis of therapeutics, 8th edn. Pergamon, New York, pp 131–149

Teicher BA, Herman TS, Holden SA, Eder JP (1991) Chemotherapeutic potentiation through interaction at the level of DNA. In: Chou T-C, Rideout DC (ed) Synergism and antagonism in chemotherapy. Academic Press, San Diego, pp 541–583

Trebst A (1987) The three-dimensional structure of the herbicide binding niche on the reaction center polypeptides of photosystem II. Z Naturforsch C 42: 742–750

Tsai C-M, Gazdar AF, Venzon DJ, Steinberg SM, Dedrick RL, Mulshine JL, Kramer BS (1989) Lack of in vitro synergy between etoposide and cis-diammine-dichloroplatinum (II). Cancer Res 49: 2390–2397

Tverskoy M, Fleyshman G, Bradley EL Jr, Kissin I (1988) Midazolam-thiopental anesthetic interaction in patients. Anesth Analg 67: 342–345

–   Ben-Shlomo I, Ezry J, Finger J, Fleyshman G (1989) Midazolam acts synergistically with methohexitone for induction of anaesthesia. Br J Anaesth 63: 109–112

Undem BJ, Adams III GK (1988) An analysis of the functional interactions of selected contractile agonists in the guinea pig isolated trachea. J Pharmacol Exp Ther 246: 47–53

Unkelbach H-D, Pöch G (1988) Comparison of independence and additivity in drug combinations. Arzneimittelforschung 38: 1–6

–   Wolf T (1984) Drug combinations – concepts and terminology. Arzneimittelforschung 34: 935–938

–   – (1985a) Qualitative Dosis-Wirkungs-Analysen. Einzelsubstanzen und Kombinationen. G Fischer, Stuttgart New York

–   – (1985b) Dose-response analysis of combination preparations. Statist Med 4: 77–85

Van den Brink FG (1973) The model of functional interaction I. Development and first check of a new model of functional synergism and antagonism. Eur J Pharmacol 22: 270–278

–   (1977) General theory of drug-receptor interactions. In: Van Rossum JM (ed) Kinetics of drug action. Springer, Berlin Heidelberg New York, pp 170–254 (Handbook of experimental pharmacology, vol 47)

Van Deventer GM, Schneidman D, Walsh JH (1985) Sucralfate and cimetidine as single agents and in combination for treatment of active duodenal ulcers. Am J Med 79 [Suppl 2C]: 39–44

Vouk VB, Butler GC, Upton AC, Parke DV, Asher SC (eds) (1987) Methods for assessing the effects of mixtures of chemicals. Wiley, Chichester

Waller DP, Katz NL, Morris RW (1980) Potentiation of lethality in mice by combinations of pentazocine and tripelennamine. Clin Toxicol 16: 17–23

Walter SD, Holford TR (1978) Additive, multiplicative, and other models for disease risks. J Epidemiol 108: 341–346

Ward WD (1988) When does synergism exist? The role of the exposure-equivalent principle. In: Manninen O (ed) Recent advances on combined effects of environmental factors. ACES, Tampere, Finland, pp 51–64

Webb JL (1963) Effects of more than one inhibitor. In: Webb JL (ed) Enzyme and metabolic inhibitors, vol 1. Academic Press, New York, pp 487–512

Weinberger MH (1985) Blood pressure and metabolic responses to hydrochlorothiazide, captopril and the combination in black and white mild-to-moderate hypertensive patients. J Cardiovasc Pharmacol 7: S52–55

Westfall DP (1981) Supersensitivity of smooth muscle. In: Bülbring E, Brading AF, Jones AW, Tomita T (eds) Smooth muscle. An assessment of current knowledge. Edward Arnold, London, pp 285–309

Wilder (1967) Stimulus and response. The law of initial value. Wright, Bristol

Winter JC (1974) Propranolol and morphine: a lethal interaction. Arch Int Pharmacodyn 212: 195–198

Wong JT-F (1975) Kinetics of enzyme mechanisms. Academic Press, London

Yajima T, Suzuki T, Suzuki Y (1988) Synergism between calcium-mediated and cyclic AMP-mediated activation of chloride secretion in isolated guinea pig distal colon. Jpn J Physiol 38: 427–443

Yamaji Y, Natsumeda Y, Yamada Y, Irino S, Weber G (1990) Synergistic action of tiazofurin and retinoic acid on differentiation of HL-60 leukemia cells. Life Sci 46: 435–42

Yonetani T, Theorell H (1964) Studies on liver alcohol dehydrogenase complexes. III. Multiple inhibition kinetics in the presence of two competitive inhibitors. Arch Biochem Biophys 106: 243–251

Zaider M (1991) The interaction of two agents revisited: response to M.C. Berenbaum. Radiation Res 126: 266–268

Subject index